Geotechnical Engineering: Pile Design and Construction

Geotechnical Engineering: Pile Design and Construction

Contributors

Jing Ma et al.

AURIS
Reference

www.aurisreference.com

Geotechnical Engineering: Pile Design and Construction

Contributors: Jing Ma et al.

Published by Auris Reference Limited

www.aurisreference.com

United Kingdom

Geotechnical Engineering: Pile Design and Construction

ISBN: 978-1-78154-914-8

British Library Cataloguing in Publication Data
A CIP record for this book is available from the British Library

Printed in the United Kingdom

Exclusively distributed by CBS Publishers & Distributors Pvt. Ltd.

Sales & Distribution Rights only for India, Pakistan, Bangladesh, Sri Lanka, Nepal and Bhutan.This book is not to be sold outside these territories.

Contents

List of Abbreviations

ANN	Artificial neural networks
ACIP	augured cast-in-place
BBN	Bayesian Belief Network
BOTDR	Brillouin Optical Time Domain Reflectometry
BHC	Butler–Hoy Criterion
CFG	Cement fly ash gravel
CKE	Chin–Kondner extrapolation
CPT	Cone Penetration Test
CFA	continuous flight auger
DOL	Davisson Offset Limit
DTL3	Down Town Line Stage 3
ERSS	Earth retaining and stabilizing structures
FBG	Fiber Bragg grating
FE	Finite element
Ge-doped	Germanium-doped
HMW	Hourly mean wind
OA	Old alluvium
PHC	Prestress high concrete
PDFs	Probability density functions
QA	quality assurance
ROC	Receiver operating characteristic
RMSE	Root mean squared error
SSI	Soil–structure interaction
TPW	Tampines West

List of Contributors

Jing Ma
School of Civil Engineering & Architecture, Chongqing Jiaotong University, Chongqing, China

Shinya Inazumi
Kyoto University, Japan

Jihui Ding
College of Civil Engineering, Hebei University, Baoding, China

Yanliang Cao
College of Civil Engineering, Hebei University, Baoding, China

Weiyu Wang
Hebei Academy of Building Research, Shijiazhuang, China

Tuo Zhao
Hebei Academy of Building Research, Shijiazhuang, China

Junhui Feng
China Metallurgical Design and Research Institute Co., Ltd., Baoding, China

Wael N. Abd Elsamee
Faculty of Engineering, Sinai University, El Arish, Egypt

Musab Aied Qissab
Department of Civil Engineering, Al-Nahrain University, Baghdad, Iraq

Fadi Hage Chehade
Civil Engineering Department, Doctotral School of Sciences and Technology, Lebanese University, University Institute of Technology (Saida) & Modeling Center, Beirut, Lebanon

Marwan Sadek
Laboratory of Civil Engineering and GeoEnvironment, University of Lille I Sciences and Technology, Villeneuve d'Ascq, France

Douaa Bachir
Numerical Center, Doctoral School of Science and Technology, Lebanese University, Beirut, Lebanon

A. M. J. Mens
Deltares, Unit Geo-Engineering, 2600 MH Delft, The Netherlands
Faculty of Civil Engineering and Earth Sciences, Delft University of Technology, Delft, The Netherlands

M. Korff
Deltares, Unit Geo-Engineering, 2600 MH Delft, The Netherlands
Cambridge University, Cambridge, UK

A. F. van Tol
Deltares, Unit Geo-Engineering, 2600 MH Delft, The Netherlands
Faculty of Civil Engineering and Earth Sciences, Delft University of Technology, Delft, The Netherlands

A. W. Stuedlein
School of Civil and Construction Engineering, Oregon State University, 101 Kearney Hall, Corvallis, OR 97331, USA

S. C. Reddy
School of Civil and Construction Engineering, Oregon State University, 101 Kearney Hall, Corvallis, OR 97331, USA

T. M. Evans
School of Civil and Construction Engineering, Oregon State University, 101 Kearney Hall, Corvallis, OR 97331, USA

C.G. Chinnaswamy
Meinhardt Infrastructure Pte Ltd, Singapore, Singapore

David N.G. Chew Chiat
Meinhardt Infrastructure Pte Ltd, Singapore, Singapore

B. R. Jayalekshmi
National Institute of Technology Karnataka, Surathkal, Karnataka, India

S. V. Jisha
National Institute of Technology Karnataka, Surathkal, Karnataka, India

R. Shivashankar
National Institute of Technology Karnataka, Surathkal, Karnataka, India

Xiaolin Weng
Key Laboratory for Special Area Highway Engineering of Ministry of Education, Chang'an University, Xi'an, Shaanxi, China

Jianxun Chen
School of Highway, Chang'an University, Xi'an 710064, China

Jun Wang
China Railway First Survey and Design Institute Group Ltd., Xi'an 710064, China

Jinxing Lai
Shaanxi Provincial Major Laboratory for Highway Bridge & Tunnel, Chang'an University, Xi'an 710064, China
School of Highway, Chang'an University, Xi'an 710064, China

Houquan Liu
School of Highway, Chang'an University, Xi'an 710064, China

Junling Qiu
School of Highway, Chang'an University, Xi'an 710064, China

Jianxun Chen
Shaanxi Provincial Major Laboratory for Highway Bridge & Tunnel, Chang'an University, Xi'an 710064, China

Yongjei Lee
Port and Harbor Team, Seoyeong Engineering, Republic of Korea

Sungchil Lee
Department of Computer Design, School of Engineering and Agriculture, Ulaanbaatar University, Mongolia

Hun-Kyun Bae
Department of Global Environment, School of Environment, Keimyung University, 203 Osan Hall, Dalgubul-Daero, Dalsegu, Daegu 1095, Republic of Korea

Preface

Geotechnical engineering is the branch of civil engineering concerned with the engineering behavior of earth materials. Geotechnical engineering is important in civil engineering, but also has applications in military, mining, petroleum and other engineering disciplines that are concerned with construction occurring on the surface or within the ground. The text *Geotechnical Engineering: Pile Design and Construction* covers construction and design aspects of piling. Influence analysis of a new building to the bridge pile foundation construction has been presented in first chapter. In second chapter, an evaluation method that can express the local leakage of leachate from the joint sections in the steel pipe sheet piles (SPSP) cutoff walls has been discussed. Experimental study of dynamic characteristics on composite foundation with CFG long pile and rammed cement-soil short pile has been focused in third chapter. A new method for prediction pile capacity executed by continuous flight auger (CFA) has been introduced in fourth chapter. In fifth chapter, the flexural behavior of laterally loaded tapered piles in cohesive soils is investigated. In sixth chapter, we present a numerical modeling of the interaction of using FLAC3D software. In seventh chapter, the application of multilayer neural network (MBPNN) as a structural analyzer for jetty structures has been explored. Wind load analysis of tall chimneys with piled raft foundation considering the flexibility of soil has been carried out in eighth chapter. Ninth chapter discusses the analysis and methodology to assess the effect on the pile foundation of a high-rise building due to the deep excavation of the down town line stage 3 (DTL3) tampines west (TPW) station. Tenth chapter focuses on the differences in the interpreted failure load for augered cast in place (ACIP) piles and seeks to determine which methods are suitable and which methods are inappropriate for the interpretation of ACIP piles. Eleventh chapter outlines a centrifugal model test, performed using a 60 g ton geocentrifuge, to investigate the performance of pipe piles used to reinforce the loess foundation below a widened embankment. The aim of twelfth chapter is to investigate the settlement behaviors of saturated tailings dam soft ground under cement fly ash gravel (CFG) pile composite foundation treatment. Last chapter introduces the receiver operating characteristic (ROC)-curve technique to estimate mainly the quality of a model and to be able to optimize parameters and variables in the model.

Chapter 1

INFLUENCE ANALYSIS OF A NEW BUILDING TO THE BRIDGE PILE FOUNDATION CONSTRUCTION

Jing Ma

School of Civil Engineering & Architecture, Chongqing Jiaotong University, Chongqing, China

ABSTRACT

This paper is based on the analysis of an industrial factory building to the bridge pile foundation construction stability, and it researches the influence of a new building to the bridge pile foundation internal force by the finite element analysis software ANSYS. By calculating the changes of displacement and internal force of the bridge pile foundation, the deformation can be better controlled. Furthermore, comparing the data of numerical analysis with one of monitor measurements, we conclude that a new building has a small influence on the deformation under load action and the stress variation of a bridge pile foundation. That is to say, the bridge pile foundation is safe and stable under load action.

INTRODUCTION

Since the capital construction increasingly develops and improves in China, more and more new buildings are built on their neighboring existing buildings [1] [2] , which have a certain influence on existing buildings. All these situations, including a foreign-style house on the shallow tunnel, a tunnel under high-rise construction, or a deep foundation ditch around the bridge [3] , require a strict computational analysis to provide reliable data for the influence extent of new buildings to existing ones and estimate the force change of building structure.

ENGINEERING SITUATION

The new industrial factory is located on a high slope, part of the tectonic denudation hilly topography. According to the original relief map, the terrain is flat in the lows, with a gradient of 35. And the slope is a little steep, with a gradient of 15 or 20. Due to a consequent bedding rock landslide, a 25-meter-high and 30-meter-long fill slope is formed on the section of 10'-10' - 15'-15', whose interface obliquity is about 20 degree, consistent with the dip angle of rock stratum.

Currently, a support reinforcement has been applied to the slope by a pile sheet wall. The length of the slope retaining wall is 587.54 meters. Fifty piles are arranged in the middle of the slope, including the bridge pile foundation support and bolt structure beam protection. Specific plans are shown in Figure 1.

ANALYSIS OF THE FINITE ELEMENT MODEL

Computation Module

To reduce the boundary effect and guarantee the accuracy in computation, the model size is that: length along slope to the factory building (X-direction) is 120 m. Width along slope to Y-direction is 50 m. Height from the lower boundary to the surface (Y-direction) is 58 m.

The whole computation module is simulated with a total of 56,326 planar units and 10,659 nodes in the finite element grid. And the finite element grid is divided as Figure 2.

Design Conditions

The model is calculated and analyzed by using Drucker-Prager Yield Criterion in ANSYS [4] , and material parameters are determined based on data from geological survey report. The results are shown in Table 1.

The finite element simulation is computed under the load of self-weight stress and additional stress respectively. We divide the jump into two phases:

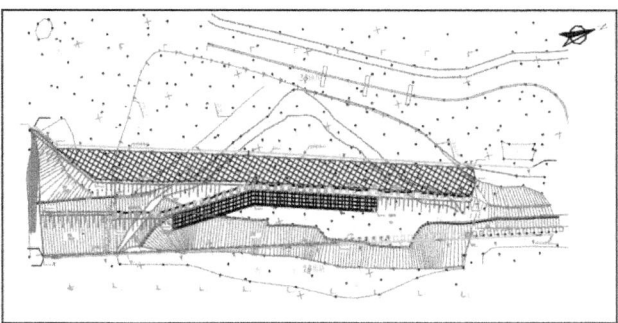

Figure 1: Master plan.

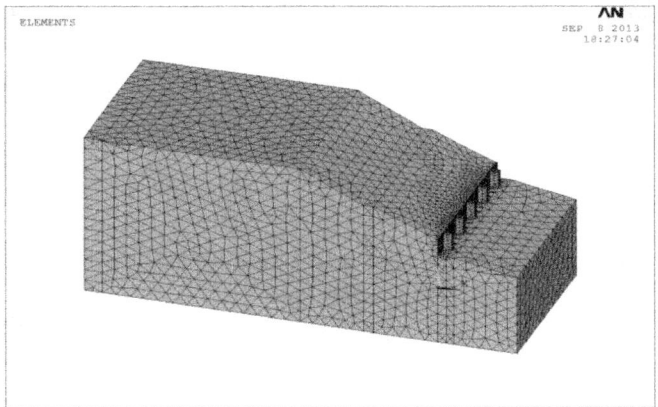

Figure 2: The finite element computation and analysis module.

Step 1: self-weight stress loading;

Step 2: factory loading.

Because the factory loading in process is subject to banded model, these loads are equivalently applied to the whole area of industrial buildings in the worst situation, and the force is 250 KN/M³. Results are shown in Figure 3.

Results

Simulation of the Results of Self-Weight Stress to the Bridge Pile Foundation

Maximum displacement and stress values in all directions of the bridge pile foundation under self-weight stress are shown in Table 2.

The nephogram of maximum displacement and stress values in all directions of the bridge pile foundation under self-weight stress are shown in Figures 4-9.

Under self-weight stress, the displacement and stress values of the bridge pile foundation are both small. The maximum values of displacement and stress of the bridge pile foundation are both in Y-direction, while the ones are small in Z-direction.

Simulation of the Results of Load Action to the Bridge Pile Foundation

Maximum displacement and stress values in all directions of the bridge pile foundation under load action are shown in Table 3.

The nephogram of maximum displacement and stress values in all directions of the bridge pile foundation under load action are shown in Figures 10-15.

Under load action, the values of displacement and stress in each direction increase, especially in Y-direction, which is consistent with reality. The results verify the correctness of the simulation.

Then, the stresses of bridge pile foundation are less than concrete compression strength in each direction, and

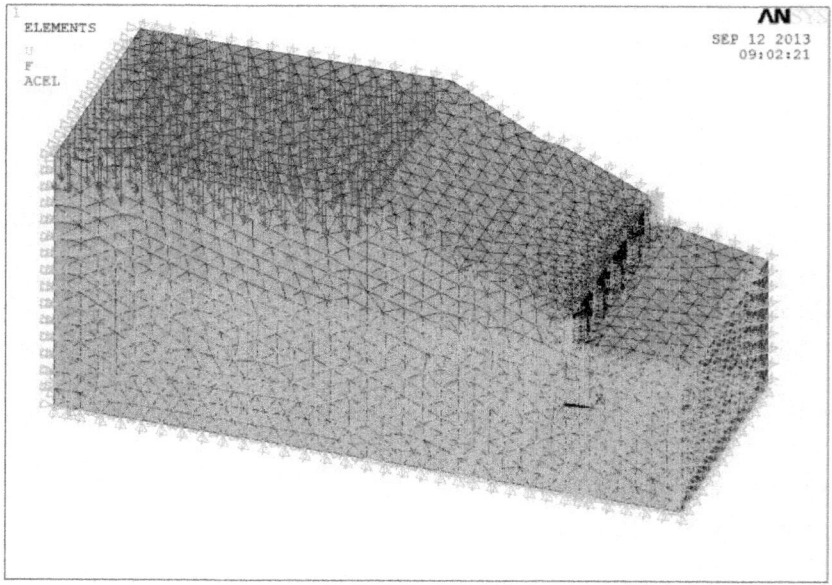

Figure 3: The loading model.

Table 1: Physical property parameter of material

Name	Multiplicity γ kg/m³	Internal frictional angle $\phi/°$	Elasticity Modulus E/GPa	Poisson ratio μ
Backfill	2000	28	1.7e−3	0.35
Mudstone	2200	35	0.2	0.23
Concrete	2400	—	30	0.20

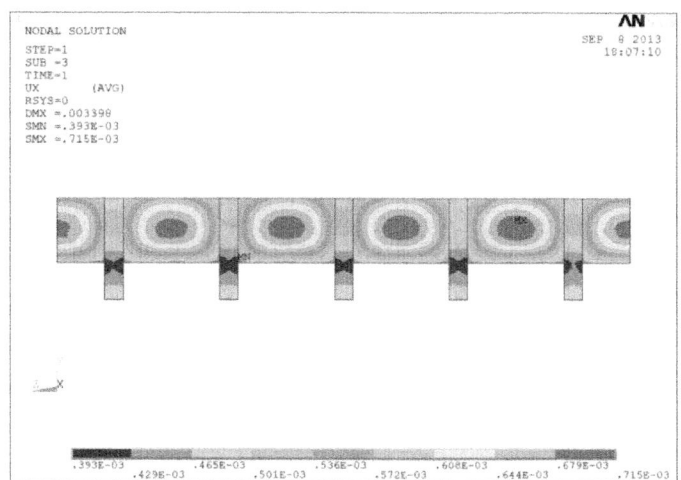

Figure 4: Displacement diagram under self-weight stress in X-direction.

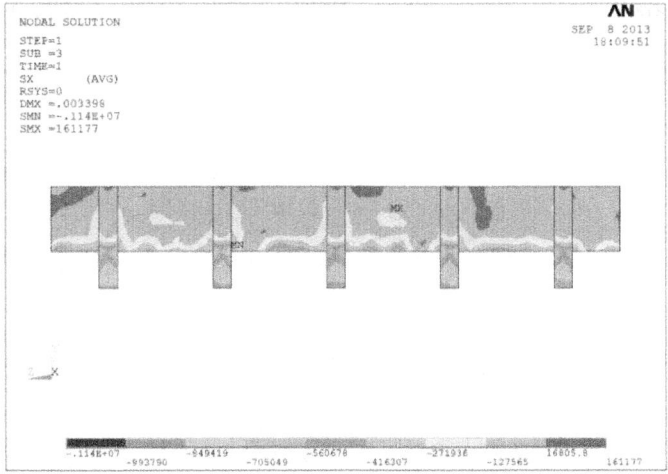

Figure 5: Stress diagram under self-weight stress in X-direction.

Table 2: Bridge pile foundation displacement and stress values under self-weight stress

		X-direction	Y-direction	Z-direction
	Displacement (mm)	0.715	−3.347	−0.345
Stress (MPa)	Maximum	0.161	0.070	0.568
	Minimum	−1.140	−1.730	−0.730

the displacements under load action all meet the load bearing requirements, which makes it reasonable and feasible to an build an industrial factory near this bridge pile foundation.

RELATIVE ANALYSIS OF THE NUMERICAL COMPUTATION AND MONITOR MEASUREMENT

Since the only data we can measure is the bridge pile foundation deformation, we set up three stations along the

Table 3: Bridge pile foundation displacement and stress values under load action

		X-direction	Y-direction	Z-direction
	Displacement (mm)	0.748	−3.354	−0.360
Stress (MPa)	Maximum	0.265	0.720	0.596
	Minimum	−1.160	−1.750	−1.190

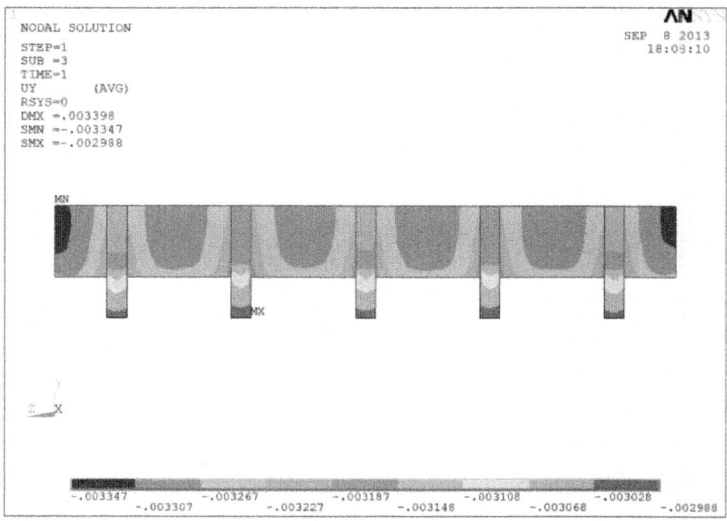

Figure 6: Displacement diagram under self-weight stress in Y-direction.

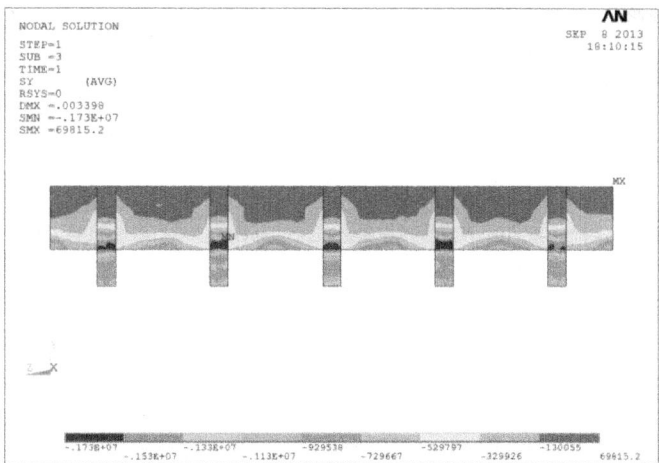

Figure 7: stress diagram under self-weight stress in Y-direction.

bridge pile foundation and conducted a long-term monitoring. We got primary data and current data of the bridge pile foundation, before and after setting up the industrial factory respectively. And the average value of three stations is chose as computed displacement value increment [5] . We compare the data of numerical simulation with that of monitor measurement to verify the reliability of the numerical simulation. Computed and measured values are shown in Table 4.

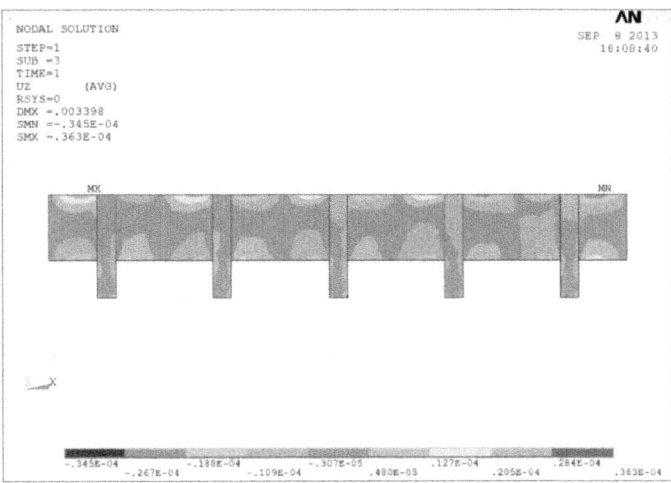

Figure 8: Displacement diagram under self-weight stress in Z-direction.

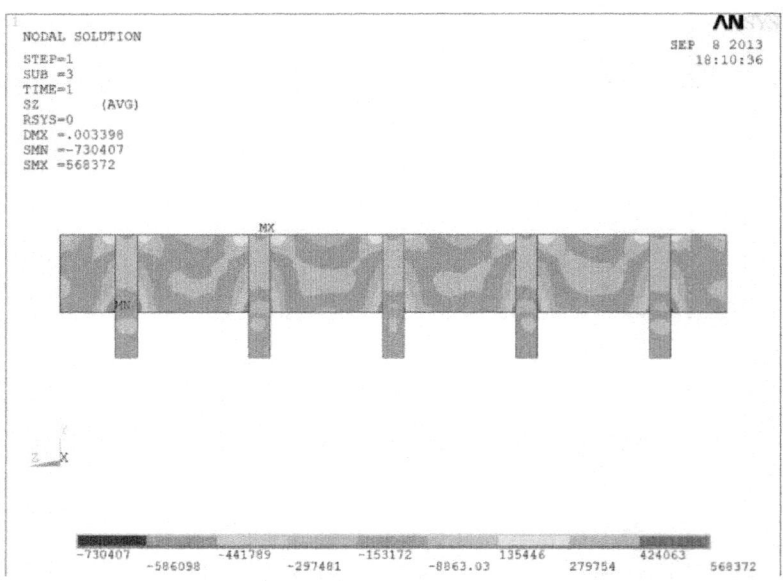

Figure 9. Stress diagram under self-weight stress in Z-direction.

Table 4: Bridge pile foundation displacement and stress values under load action

		X-direction	Y-direction	Z-direction
Computed stress (MPa)	Maximum increment	0.104	0.650	0.028
	Minimum increment	−0.020	−0.020	−0.460
Computed displacement value increment (m)		0.033	−0.007	−0.091
Measured displacement value (m)		0.030	−0.005	−0.082

The main displacement is that in X-direction (horizontal direction) under load action, and it differs by 0.003 m. The maximum stress change of the bridge pile foundation is 0.65 MPa in Y-direction (vertical direction), which differs by 0.002 m between computed and measured values. The value in Z-direction is perpendicular to the horizontal direction, which differs by 0.009 m between computed and measured values.

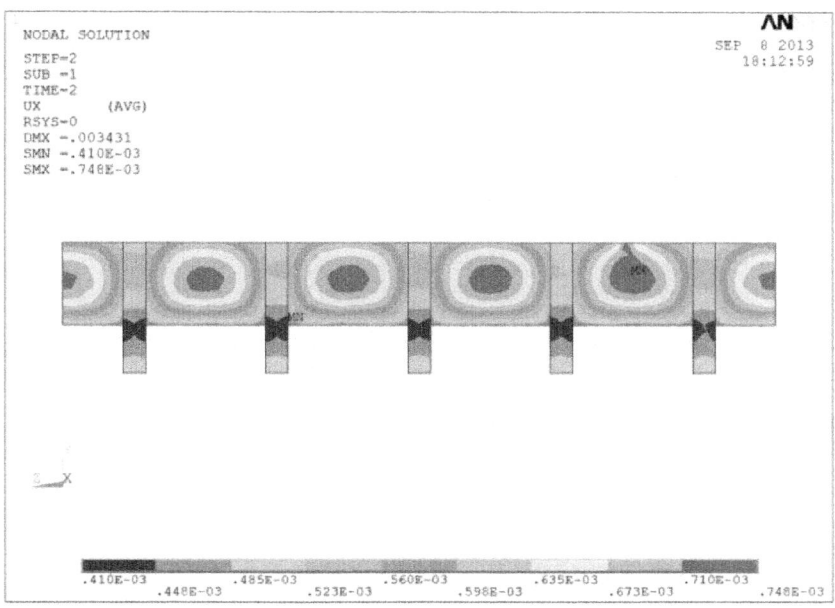

Figure 10: Displacement diagram under load action in X-direction.

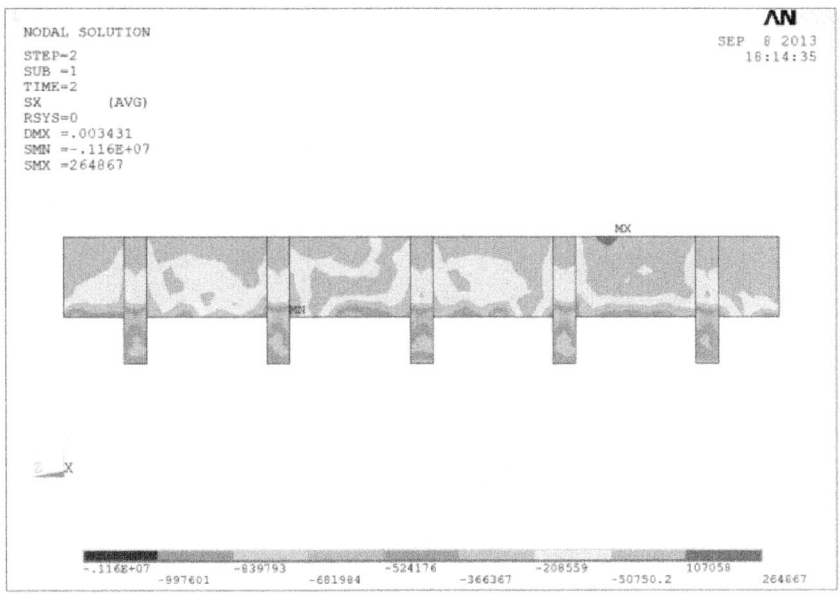

Figure 11: Stress diagram under load action in X-direction.

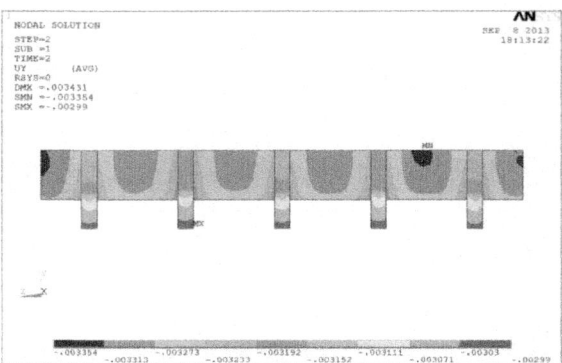

Figure 12: Displacement diagram under load action in Y-direction.

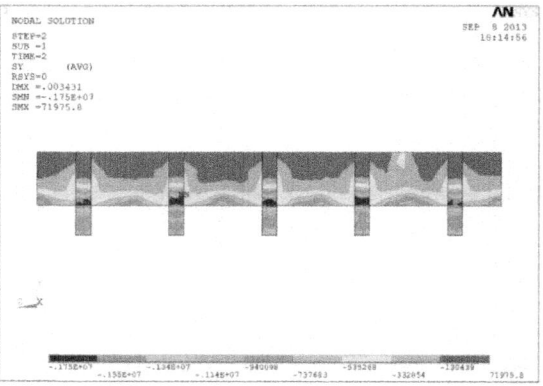

Figure 13: Stress diagram under load action in Y-direction.

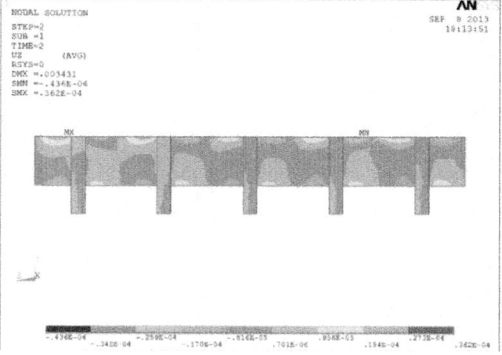

Figure 14: Displacement diagram under load action in Z-direction.

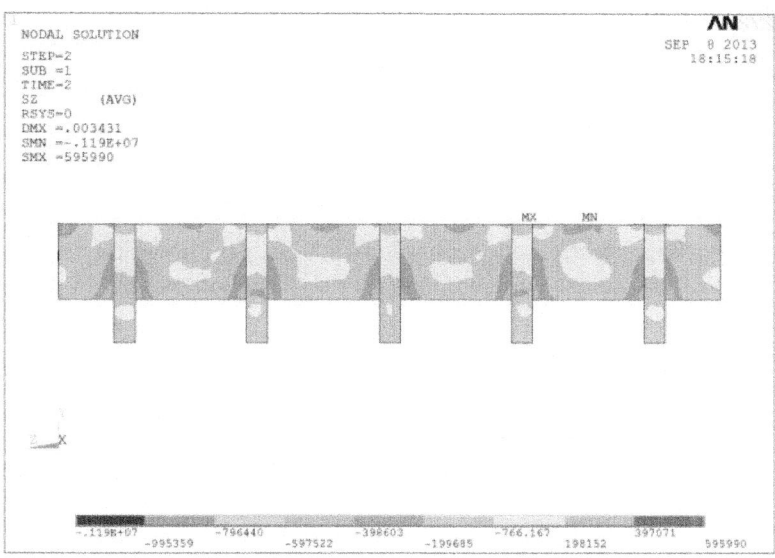

Figure 15: Stress diagram under load action in Z-direction.

CONCLUSIONS

Stress and displacement values of supporting structure are computed and analyzed by the finite element software ANSYS. The comparisons between computed and measured values are illustrated as below.

1)   There is little difference between computed and measured values in all directions of the bridge pile foundation under load action, particularly in the main displacement (X-direction), which differs by 0.003 m. It shows that the load we applied about 250 c is rational and the computed values are reliable.

2)   The maximum stress change of bridge pile foundation under load action is 0.46 MPa in Z-direction, less than concrete compression strength. Hence the bridge pile foundation meets the load bearing requirements.

3)   The maximum tensile stress change of bridge pile foundation under load action is 0.65 MPa in Y-direction (vertical direction), less than tensile strength of concrete. Hence the bridge pile foundation meets the load bearing requirements.

Analysis above illustrates that the impact on the displacement and stress change of bridge pile foundation under load action is small. The bridge pile foundation structure under load action is safe and stable.

REFERENCES

1.   Liu, J.H. (2002) Several Theories and Calculated Methods of Underground Engineer Construction Mechanics. Railway Standard Design.

2.   Hu, B. (2013) Application of Reinforced Concrete Filling Pile Technology in Deep Foundation Pit Supporting. Shandong Metallurgy, 35, 74-75.

3.   Wang, Y.G., Li, D.J. and Ye, K.P. (2005) Application of Steel-Tube Supporting Prestressed Technology in Deep Foundation Pit Project in Shanghai Xianlesi Square. Construction Technology, 35, 45-48.

4.   Mo, H.O., Crèpe, H.X. and Lai, A.P. (2001) Optimization Design in Foundation Pit Supporting Pile Structure. Rock Mechanics and Engineering, 23, 23-25.

5.   Paw-paw, H.Y. and Huang, J.Z. (2001) 3D Finite Element Analysis and Simulation of Deep Foundation Pit Supporting Structure. Shanghai Jiaotong University, 35, 610-613.

Chapter 2

HYDRAULIC CONDUCTIVITY OF STEEL PIPE SHEET PILE CUTOFF WALLS AT COASTAL WASTE LANDFILL SITES

Shinya Inazumi

Kyoto University, Japan

INTRODUCTION

Landfill sites are facilities where the final residue is disposed after all possible recycling energy has been recovered from it. Therefore, landfill sites are an important part of civil infrastructure, required for environmental conservation without dumping waste in residential areas. However, in many cases, the construction of landfill sites has been opposed due to concerns of residents living the vicinity regarding environment safety with regard to situations such as "the leachate from waste may leak out"; hence, the construction of new landfill sites has become more difficult. Moreover, the construction cost of landfill sites has also significantly increased simultaneously due to tighter environmental legislation (Shimizu, 2003; Kamon et al., 2007).

In Japan, small-scale inland landfill sites were often constructed in the river-head areas of mountain valleys. With regard to the abovementioned social concerns regarding the landfill sites, the locations of landfills have recently been diversified into coastal areas on a large scale. These sites are developed at urban harbour areas in order to reduce the risk of contaminating the groundwater, which can be caused by the leakage of leachate, and conserve the water resources (Kamon & Inui, 2002). In the national statistics of 2003 announced at Ministry of the Environment, the capacity of coastal landfill sites was 23.3% of that of all landfill sites, and particularly in metropolitan areas, it was greater than 80% (seeFig. 1). These statistics indicate that the role of coastal landfill sites has been increasing steadily. However, the residents living in the vicinity of these sites continue to express the same concerns for environment safety. Therefore, ensuring stable and systematic operation of the coastal landfill sites in the future and prolonging the life of coastal landfill sites constructed until now are important matters of concern, particularly in metropolitan areas.

A revetment at a coastal landfill site ensures space for waste disposal and harbour maintenance during the disposal of waste, construction sludge, dredged soil etc. A revetment at a coastal landfill site must function as a vertical (side) cutoff barrier that prevents the leakage of leachate containing toxic substances from the landfill waste, into the sea; furthermore revetments must protect the coastal landfill site from various external forces such as earthquakes, ocean waves, high tides and tsunamis (Waterfront Vitalization and Environment Research Center, 2002).

Recently, steel pipe sheet piles (SPSPs), using which the deepwater construction is possible (Japanese Association for Steel Pipe Piles, 1999), have been widely employed in vertical cutoff barriers at coastal landfill sites due to their workability and economical efficiency.

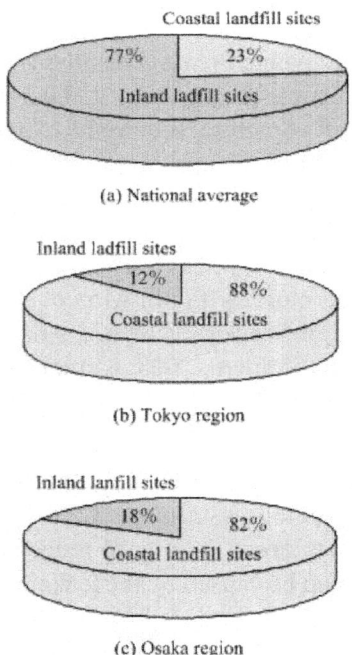

Coastal landfill sites

77%   23%

Inland ladfill sites

(a) National average

Inland ladfill sites

12%   88%

Coastal landfill sites

(b) Tokyo region

Inland lanfill sites

18%   82%

Coastal landfill sites

(c) Osaka region

Figure 1: Capacity comparison between inland and coastal landfill sites based on national statistics of 2003 announced at Ministry of the Environment, Japan itle of figure, left justified

Avertical cutoff barrier employing SPSPs is called a "SPSP cutoff wall" in this study. However, the design and application of SPSP cutoff walls, evaluation of environmental feasibility, construction technology and long-term maintenance are very complicated both experimentally and analytically (Kamon et al., 2001). This is because of the existence of joint sections in the SPSPs, as shown in Fig. 2.The appropriately estimation of the hydraulic performance of SPSPs with joint sections (shown in Fig. 2) is an important issue, particularly in the evaluation of environmental feasibility, that is, the containment of leachates containing toxic substances. Figure 3 shows the characterization of the environmental feasibility of vertical and bottom cutoff barriers as well as the overall landfill site. When evaluating the hydraulic performance of an SPSP cutoff wall, an equivalent hydraulic conductivity is generally obtained (Waterfront Vitalization and Environment Research Center, 2002). This equivalent hydraulic conductivity assumes that the joint section and the steel pipe are integrated; therefore, the hydraulic conductivity is substituted with a uniform permeable layer (see Fig. 4). The Prime Minister's Office and the Ministry of Health and Welfare says that the integrated equivalent hydraulic conductivity with 50 cm thickness must be 1.0×10^{-6} cm/s or less (Waterfront Vitalization and Environment Research Center, 2002). However, in an evaluation that employs the equivalent hydraulic conductivity, it is difficult to consider the local leakage of leachate containing toxic substances from the joint sections in the SPSP cutoff wall.

In this study, an evaluation method that can express the local leakage of leachate from the joint sections in the SPSP cutoff walls is discussed. In particular, the evaluation of the environmental feasibility (containment of leachates containing toxic substances) considering a

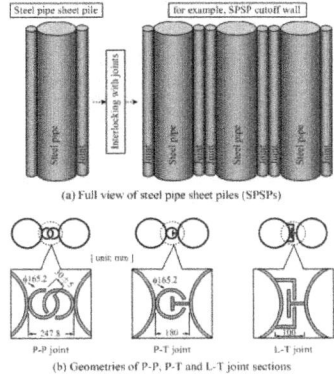

Figure 2: Schematic diagram of steel pipe sheet piles with joint sections

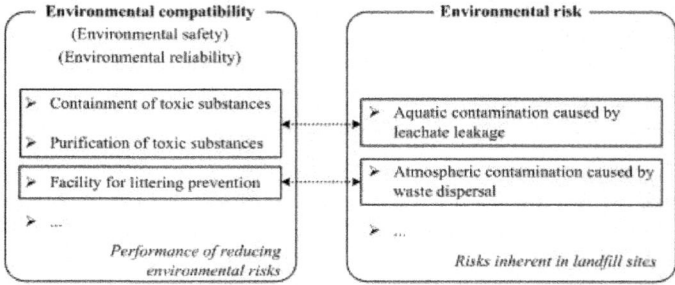

Figure 3: Characterization of environmental feasibility on vertical and bottom cutoff barriers as well as overall landfill site

three-dimensional arrangement and hydraulic conductivity distribution of the joint sections in the SPSP cutoff wall is compared with an evaluation that uses the equivalent hydraulic conductivity.

ANALYSIS FOR ENVIRONMENTAL FEASIBILITY

The development of methods for the detection of the generation points of leachate leakage has been conducted in various different ways at inland and coastal landfill sites in order to determine when the leachate containing toxic substances will leak into the surrounding areas after the land has been reclaimed at the landfill site (Kamon & Jang, 2001; The Landfill System & Technologies Research Association of Japan, 2004). However, the present detection methods are insufficient with regard to their durability, and the use of these methods may lead to excess cost and time for repairing the generation points of leachate

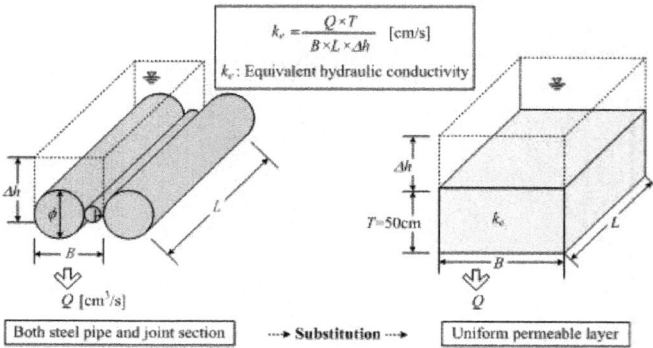

Figure 4: Concept of equivalent hydraulic conductivity assuming that joint section and steel pipe are integrated

leakage in the vertical and bottom cutoff barriers at the landfill sites. Therefore, an effective implementation and verification of the seepage and advection/dispersion analysis, considered as a two-dimensional or a three-dimensional problem, of the leaching behavior of leachate containing toxic substances are necessary along with the upgradation of the technique used to repair vertical and bottom cutoff barriers. The structure of vertical and bottom cutoff barriers that can ensure long-term stability as well as the evaluation method for the environmental feasibility of landfill sites must be also discussed.

The leaching behavior of leachates containing toxic substances near the vertical and bottom cutoff barriers at landfill sites must be considered with regard to not only infiltration but also the advection and dispersion phenomena (Kamon et al., 2007). Therefore, these phenomena must be accurately reproduced in the implementation of the seepage and advection/dispersion analysis. In this study, the infiltration, advection and dispersion phenomena must be expressed three-dimensionally in order to account for the joint sections in the SPSP cutoff walls. Also, the analysis of coastal landfill sites, unlike that for inland landfill sites, must consider the effect of tides, etc. Furthermore, each vertical and bottom cutoff barrier is a composite structure consisting of synthetic fiber, steel, rubble and the seabed; this composite structure must be reproduced accurately.

The Eulerian-Lagrangean finite-element method is a numerical calculation method that is known to be useful in efficiently reproducing such complicated phenomena. In this study, the seepage and advection/dispersion analysis is performed using Dtransu-3D/EL, which is used as a representative analysis code (Nishigaki et al., 1995).

Objective and Assessment Index

In an SPSP cutoff wall, joint sections are arranged between steel pipes, forming a three-dimensional structure (see Fig. 2). Therefore, it is necessary to accurately reproduce the local leakage of leachates from the joint sections for the evaluation of the environmental feasibility of the SPSP cutoff wall. In this study, the leachate-containment effect of the SPSP cutoff wall is evaluated by using a three-dimensional seepage and advection/dispersion analysis (Dtransu-3D/EL). This analysis reproduces the existence of joint sections more precisely.

Figure 5 shows the three-dimensional cross-section of a landfill site assumed as a basic case in this analysis. The SPSP cutoff wall as well as a part of the composition layer around it in

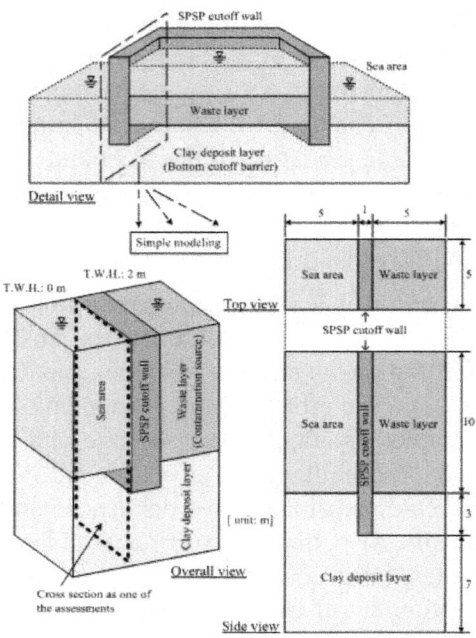

Figure 5: Three-dimensional cross section of landfill site assumed as a basic case in the analysis

the coastal landfill site is considered for setting the three-dimensional cross-section. At the bottom of the waste layer as well as in the sea bed, a clay deposit layer is assumed to exist, and this layer fulfils the role as a bottom cutoff barrier in the coastal landfill site. The SPSP cutoff wall is penetrated upto a depth of 3 m in the clay deposit layer, and the hydraulic conductivity of the SPSP cutoff wall is varied to provide different examination cases.

In the construction of the SPSP cutoff wall at coastal landfill sites, double SPSP cutoff walls may be used due to ensure mechanical stability and fail-safe concept of landfill sites, as shown in the overview in Fig. 6. Furthermore, the clay deposit layer may be improved by sand compaction pile (SCP) methods in order to enhance the mechanical stability of the SPSP cutoff walls (Waterfront Vitalization and Environment Research Center, 2002). However, the main objective of this study is the evaluation of the environmental feasibility (containment effect of leachate containing toxic substances) of the SPSP cutoff wall. Therefore, the coastal landfill site is simplified, as shown in Fig. 5, as a three-dimensional cross-section that comprises a single SPSP cutoff wall, waste layer and clay deposit layer. The three-dimensional cross-section assumes the extreme conditions for the vertical and bottom cutoff barriers that would pose environmental pollution risks to the surroundings affected by coastal landfill sites.

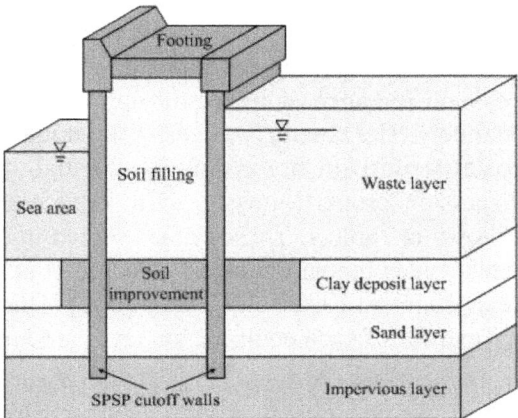

Figure 6: Overview of vertical and bottom cutoff barriers generally constructing at coastal landfill sites

Table 1: Seepage, advection and dispersion properties assigned to each composition layer in the analysis

			SPSP cutoff wall			Clay deposit layer	Waste layer	Sea area
			UL-model	SP/JS-model				
				Joint sec.	Steel pipe			
Horizontal hydraulic conductivity	k_H	cm/s	2.0×10^{-6}, 1.0×10^{-6}, 1.0×10^{-7}, 1.0×10^{-8}	2.5×10^{-6}, 1.3×10^{-6}, 1.3×10^{-7}, 1.3×10^{-8}	infinitesimal	7.0×10^{-7}	1.0×10^{-0}	1.0×10^{-0}
Vertical hydraulic conductivity	k_V	cm/s	2.0×10^{-6}, 1.0×10^{-6}, 1.0×10^{-7}, 1.0×10^{-8}	2.5×10^{-6}, 1.3×10^{-6}, 1.3×10^{-7}, 1.3×10^{-8}	infinitesimal	5.0×10^{-7}	1.0×10^{-0}	1.0×10^{-0}
Effective porosity	θ		0.1	0.1	0.1	0.65	1	1
Longitudinal dispersion	α_L	cm	10	10	infinitesimal	10	10	10
Transverse dispersion	α_T	cm	0.1	0.1	infinitesimal	1	1	1
Molecule diffusion coefficient	D_m	cm²/s	1.0×10^{-5}	1.0×10^{-5}	infinitesimal	1.0×10^{-5}	1.0×10^{-5}	1.0×10^{-5}
Retardation factor	R_d		1	1	1	2	1	1

In coastal landfill sites, the difference in the water level between the inside and outside landfill site is controlled on a daily basis so that it may not exceed 2 m (Waterfront Vitalization and Environment Research Center, 2002). On the other hand, in the three-dimensional cross-section shown in Fig. 5, a controlled water level regulated to 2 m is reproduced by the boundary conditions, that is, a fixed total head of 0 and 2 m are assigned to the upper sides of the sea area and waste layer, respectively. The boundary edges in the three-dimensional cross-section of the coastal landfill site are assumed to be undrained. The seepage, advection and dispersion properties assigned to each composition layer in this analysis are shown at Table 1. These values shown in Table 1 are typical one for heavy metals and composition layers (Kamon et al., 2001; Waterfront Vitalization and Environment Research Center, 2002). This analysis assumes that mechanical properties of each composition layer are not considered.

Presently, in Japan, waste discharge waste is burnt once at a refuse incinerator plant, and the incinerated residue generated from the incinerator plant is mainly used to reclaim land at landfill sites (Kamon & Inui, 2002). Therefore, the type of waste dumped in the recently constructed landfill sites has changed from the conventional organic substances to inorganic substances; thus, the heavy metals which may be contained in the incinerated residue are among the major environmental pollutants. If the leachate leakage occurs at a landfill site into the surrounding areas, the heavy metals also may leak out together with the leachate due to the advection-dispersion phenomenon, as heavy metals are soluble in water. Therefore, this study assumes heavy metals as toxic substances that may leak out from coastal landfill sites. This analysis assumes the waste layer to be a contamination source, and the concentration of toxic substances (heavy metals) at the waste layer is assigned the value of 1 as the initial condition. The initial relative concentration of toxic substances is initialized to 0 in regions outside the waste layer.

As an environmental conservation standard for coastal landfill sites (The Landfill System & Technologies Research Association of Japan, 2004), the environmental standard values (see Table 2 (b) and (c)) for water quality and bottom sediment of the sea areas near landfill sites equal 0.1 times that of the acceptable standard values (see Table 2(a)) for waste disposed at landfill sites. Therefore, the concentration of toxic substances at the SPSP cutoff wall on the sea side (that is the cross-section delimited by the broken line at Fig. 5) is targeted in this analysis as an important index of the environmental feasibility of SPSP cutoff walls. In this analysis, the elapsed time during which the concentration of toxic substances reaches 0.1 on the sea side of the SPSP cutoff wall is estimated; when this occurs, the SPSP cutoff wall as well as the coastal landfill site is defined as having lost its environmental feasibility.

SP/Js-Model Considering Local Water Leakage in Joint Sections

In the evaluation of the environmental feasibility (containment effect of leachate containing toxic substances) of SPSP cutoff walls at coastal landfill sites, the equivalent hydraulic conductivity is generally used (Waterfront Vitalization and Environment Research Center, 2002). This method involves calculating the hydraulic conductivity of an SPSP cutoff wall equivalent to a uniform permeable layer of thickness 50 cm (see Fig. 4) by considering the steel pipes and joint sections that constitute the SPSP cutoff wall as a single body. Because the equivalent hydraulic conductivity can be directly verified with the technical standards for vertical and bottom cutoff barriers at landfill sites, it is frequently used in the technical development of the SPSP cutoff wall. However, the value equivalent hydraulic conductivity is the average hydraulic conductivity of the joint sections, which have high permeability, and that of the steel pipe sections, which are impermeable. Therefore, an evaluation using the equivalent hydraulic conductivity cannot easily detect the position or the time of leachate leakage, thus making it difficult to estimate the environmental impact of local leakage from the joint sections of the SPSP cutoff wall. Where, development of these detections will contribute strongly for the optimization of maintenance and management in SPSP cutoff wall.

Table 2: Environmental conservation standards associated with inland and coastal landfill sites

Type of metals	Allowable limit
Cadmium and its compounds	0.1 mg/L or less
Lead and its compounds	0.1 mg/L or less
Hexavalent chromium compounds	0.5 mg/L or less
Mercury and its compounds	0.005 mg/L or less

(a) For industrial waste reclaimed in landfill sites

Type of metals	Allowable limit
Cadmium its compounds	0.01 mg/L or less
Lead and its compounds	0.01 mg/L or less
Hexavalent chromium compounds	0.05 mg/L or less
Mercury and its compounds	0.0005 mg/L or less

(b) For water pollution of groundwater

Type of metals	Allowable limit
Cadmium its compounds	0.01 mg/L or less
Lead and its compounds	0.01 mg/L or less
Hexavalent chromium compounds	0.05 mg/L or less
Mercury and its compounds	0.0005 mg/L or less

(c) For soil contamination

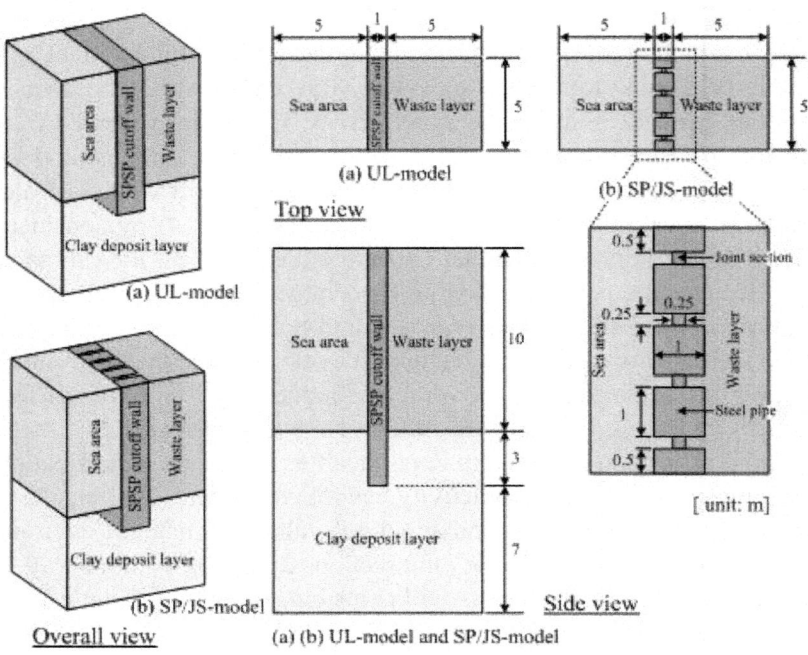

Figure7: General description of UL-model and SP/JS-model in the analysis

In this study, an evaluation method that can express the local leakage at the joint sections of SPSP cutoff walls is discussed. The evaluation method using the equivalent hydraulic conductivity is defined as the "UL-model", and the evaluation method that considers the steel pipe and joint sections, that is, the local leachate leakage, is defined as the "SP-JS-model". Figure 7 shows a general description of the UL-model and SP/JS-model. In the UL-model (shown in Fig. 7(a)), equivalent hydraulic conductivities of 2.0×10^{-6}, 1.0×10^{-6}, 1.0×10^{-7} and 1.0×10^{-8} cm/s were assigned to the entire SPSP cutoff wall. In the SP/JS-model (see Fig. 7(b)), the joint sections were placed at 0.25 m intervals for steel pipes of diameter 1 m, which represents the standard sizes of the SPSP shown in Fig. 2. Furthermore, hydraulic conductivities were assigned to each steel pipe and joint section in the SP/JS-model such that the entire hydraulic conductivity of the SPSP cutoff wall equals the equivalent hydraulic conductivity assigned in the UL-model, that is, hydraulic conductivities of 2.5×10^{-6}, 1.3×10^{-6}, 1.3×10^{-7} and 1.3×10^{-8} cm/s were assigned to the joint sections, assuming that the hydraulic conductivity of steel pipe is infinitely small.Table 1 shows the seepage, advection and dispersion properties assigned to each composition layer in both the models.

RESULTS AND DISCUSSION

Environmental Feasibility of SPSP Cutoff Wall Considering Local Water Leakage

Figure 8 shows the concentration flux (the material quantity passing through a unit area in unit time) of toxic substances leaking from the SPSP cutoff wall on the sea side. The fluxes in the uniform layer of the UL-model and in each steel pipe and joint section of the SP/JS-model are plotted in Fig. 8. The relationship between the elapsed time and the highest concentration of toxic substances leaked from the SPSP cutoff wall on the sea side for both the models is shown in Fig. 9. Figure 10 illustrates the distribution of the concentration of toxic substances leaking out from the waste layer, which is the contaminated source, for both the models. Figure 10 expresses the distribution of the concentration on the sea side of the SPSP cutoff wall in order to facilitate the comparison of both the models with regard to the leakage of the toxic substance to the surroundings of the coastal landfill site.

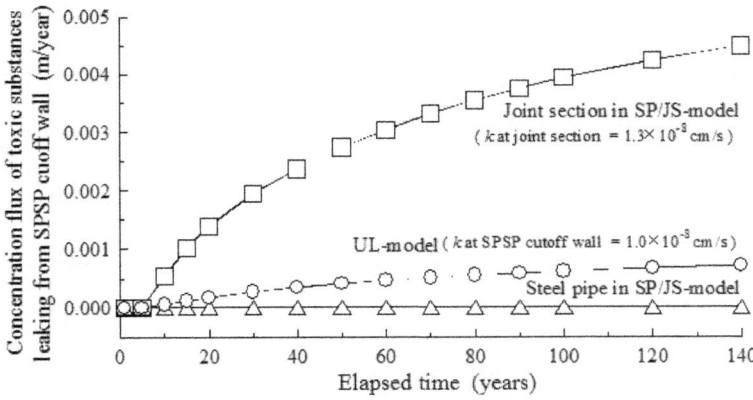

Figure 8: Concentration flux of toxic substances leaking from SPSP cutoff wall on sea side with elapsed time for both models

In the SP/JS-model, the concentration flux of toxic substances leaked onto the sea side of the SPSP cutoff wall, particularly from the joint sections, is increased as compared to that of the UL-model (seeFig. 8). The SP/JS-model can quantitatively express the concentration of toxic substances at the joint sections of the SPSP cutoff wall, where the hydraulic conductivity is higher than that in the steel pipe. In the UL-model, as shown in Fig. 10, the leachate leaks uniformly from the SPSP cutoff wall onto the sea side, and this leakage tends to uniformly increase with time. In the SP/JS-model, it being different from the UL-model, the leachate leaks locally from the joint sections onto the sea side of the SPSP cutoff wall, and this leakage increases locally with time at the joint sections (see Fig. 10). Consequently, the increase in the concentration of toxic substances leaked from the SPSP cutoff wall onto the sea side is found to occur earlier in the SP/JS-model than in the UL-model, as shown in Fig. 9.

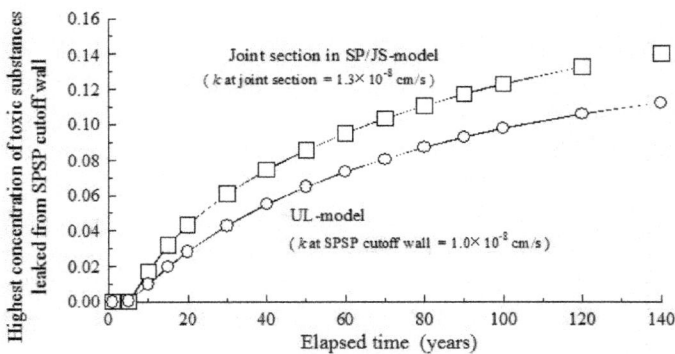

Figure 9: Relationship between elapsed time and the highest concentration of toxic substances leaked from SPSP cutoff wall on sea side for both models

For example, 70 and 110 years, respectively, are required in the SP/JS-model (the hydraulic conductivity of the entire SPSP cutoff wall is 1.0×10^{-8} cm/s) and the UL-model for the concentration of toxic substances in the SPSP cutoff wall on the sea side to reach C=0.1, which is assumed as the assessment index. In the other analyzed conditions under which the hydraulic conductivity of the entire SPSP cutoff wall is equivalent in both models, the leakage of leachate is confirmed to occur earlier in the SP/JS-model than in the UL-model due to effect of the local leakage of leachate (see Fig. 11). This tendency becomes more remarkable with increase in the hydraulic conductivity of the entire SPSP cutoff wall (see Fig. 12).

Thus, as mentioned above, the reproduction of the local leakage of leachate generated at the joint sections of SPSP cutoff walls is possible by using the SP/JS-model for the evaluation of the environmental feasibility of SPSP cutoff walls at coastal landfill sites. Furthermore, the SP/JS-model indicates that toxic substances in concentrations exceeding the environmental standard values are leaked out of coastal landfill sites earlier than that estimated using the UL-model (see Fig. 9). Using the UL-model, the local leakage of leachate containing toxic substances from the SPSP cutoff wall cannot be reproduced, although the total quantity of the toxic substances leaked from the SPSP cutoff wall can be estimated. This provides a safer-side estimate of the environmental feasibility from the viewpoint of the time taken for the leakage of toxic substances. In addition, by using the UL-model, it is difficult to quantitatively detect the generation position in the SPSP cutoff wall where the leachate containing toxic substances are leaked. An appropriate estimation in terms of both position and time at which the loss of environmental feasibility occurs is important in order to control and maintain a long-term SPSP cutoff wall at coastal landfill sites. Based on the abovementioned points, the environmental feasibility of SPSP cutoff walls must be verified by using the SP/JS-model.

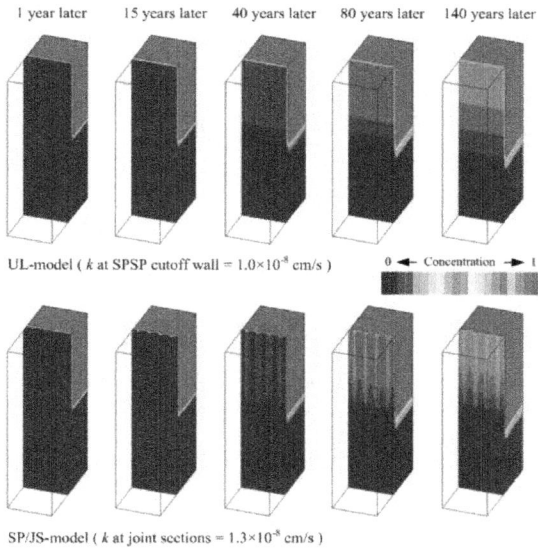

Figure 10: Distribution of concentration of toxic substances leaking out from waste layer for both models

Environmental Feasibility of SPSP Cutoff Wall Considering Joint Sections

Various types of joints are adopted for the joint sections of the SPSP cutoff walls, as shown in Fig. 2. The types of joints for which the hydraulic performance has been reported experimentally are the P-T joints in which the packing mortar is filled in the joint space, the improved P-T joint in which a rubber board is installed with the mortar filling in the joint space and the H-H joint for H-jointed SPSP in which a water-swelling sheet is applied in the joint spaces (Oki et al., 2003; Inazumi et al., 2005, 2006;Kimura et al., 2007). Based on past reports, the SPSPs with the P-T joint, improved P-T joint and H-H joint exhibit equivalent hydraulic conductivity levels of 1×10^{-6}, $1 \times 10^{-}$

[8] and 1×10^{-9} cm/s, respectively, under specific experimental conditions under which the difference between the water levels inside and outside the landfill site is less than 5 m (Oki et al., 2003; Inazumi et al., 2005, 2006). However, the reported hydraulic performances of the SPSP cutoff walls with the joint sections has been based on the equivalent hydraulic conductivities obtained from experimental studies.

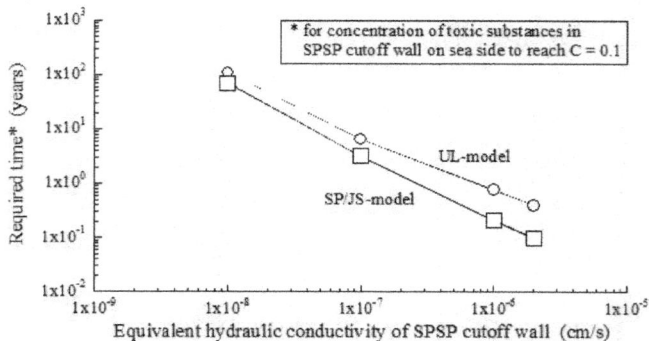

Figure 11: Required time for concentration of toxic substances in SPSP cutoff wall on sea side to reach C = 0.1 with equivalent hydraulic conductivity of SPSP cutoff wall

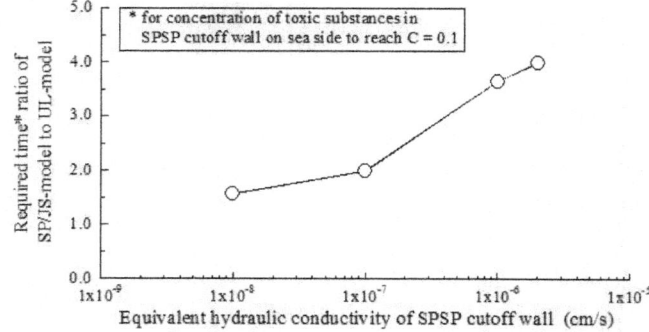

Figure 12: Required time ratio of both models, for concentration of toxic substances in SPSP cutoff wall on sea side to reach C = 0.1, with equivalent hydraulic conductivity of SPSP cutoff wall

In this study, the reported equivalent hydraulic conductivities of SPSP cutoff walls are converted to individual hydraulic conductivities in the steel pipe and joint sections.

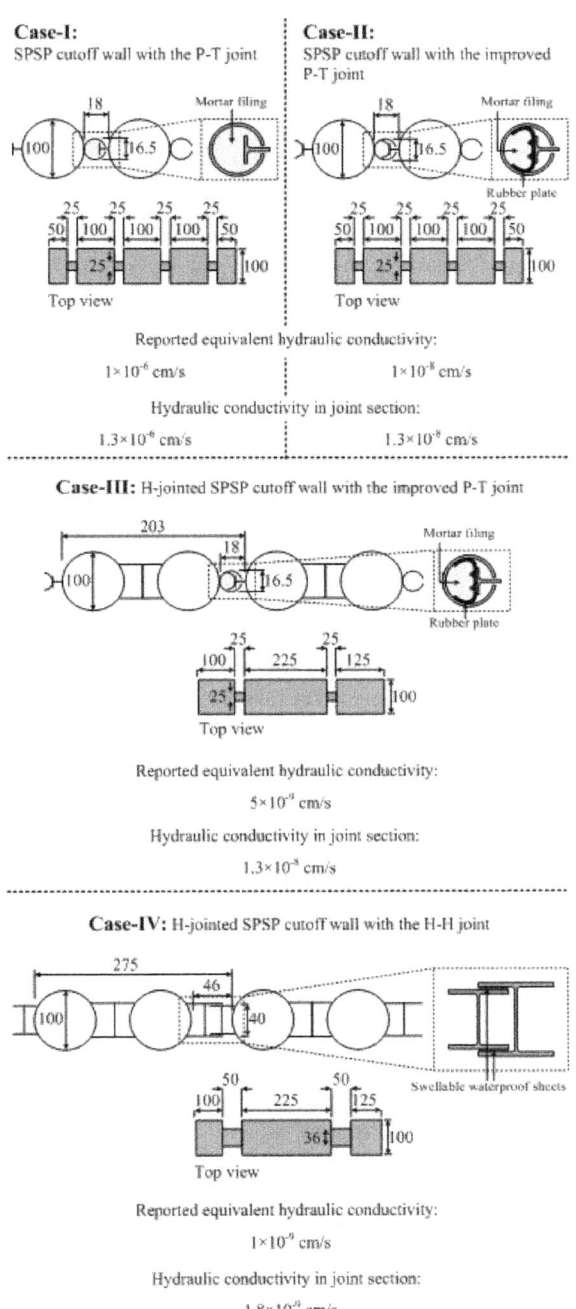

Figure 13: Dimension and hydraulic conductivity of SPSP cutoff wall with each joint type and outline of SP/JS-model for Case-I to Case-IV

Furthermore, the environmental feasibilities of SPSP cutoff walls with various joints types are evaluated by applying each converted hydraulic conductivity in the SP/JS-model. Figure 13 shows the equivalent hydraulic conductivities of SPSP cutoff walls with various joints types, the dimension of each joint type as well as steel pipe and the hydraulic conductivity of each joint type. In the evaluation of the environmental feasibilities on SPSP cutoff walls considering various joint geometries and performance levels, the SPSP cutoff walls with the following four joint types are applied to the SP/JS-model.

- Case-I:SPSP cutoff wall with the P-T joint

- Case-II:SPSP cutoff wall with the improved P-T joint

- Case-III:H-jointed SPSP cutoff wall with the improved P-T joint

- Case-IV:H-jointed SPSP cutoff wall with the H-H joint

Figure 13 shows also the outline of the SP/JS-model for Case-I to Case-IV. Joint sections of width 0.25 m and steel pipes of diameter 1 m were used in Case-I and Case-II, whereas joint sections of widths 0.25 m and 0.5 m were used in Case-III and Case-IV, respectively, along with H-jointed steel pipes of diameter 2.25 m (Oki et al., 2003; Inazumi et al., 2005, 2006). Table 1 shows the seepage, advection and dispersion properties assigned to each composition layer

The assumed hydraulic conductivities of the joint sections were 1.3×10^{-6} cm/s in Case-I, 1.3×10^{-8} cm/s in Case-II and Case-III and 1.8×10^{-9} cm/s in Case-IV.

Figure 14 shows the total quantities of toxic substances leaked from the SPSP cutoff wall onto the sea side with respect to the elapsed time for Case-I to Case-IV. The relationships between the elapsed time and the highest concentration of toxic substances leaked from the SPSP cutoff wall onto the sea side for Case-I to Case-IV are shown in Fig. 15. Figure 16 shows the distribution of the concentration of toxic substances leaking out from the waste layer, that is, the contamination source, in Case-I to Case-IV. This distribution in Fig. 16 is expressed from the sea side of the SPSP cutoff wall in order to facilitate a comparison among Case-I to Case-IV with regard to the leakage of the toxic substance to outside the coastal landfill.

The times required for the concentration levels on the sea side to exceed C=0.1 were less than 1 year and 70 years for Case-I and the Case-II, respectively (see Fig. 15). In Case-III and Case-IV, the leakage of toxic substances in excess of environmental standard value (C=0.1) was not confirmed, even for durations upto 140 years. In Case-I and Case-II, the hydraulic conductivities of the joint sections are different, although the arrangement intervals of the joint sections are the same; thus it has been proven that low-hydraulic conductivity joint sections in SPSP cutoff walls significantly contribute toward increasing the leachate-containment effect. In addition, the sparser arrangement of joint sections represented in Case-III reduces the total quantity of toxic substances leaked from the SPSP cutoff wall onto the sea side to half that in Case-II (see Fig. 14). Consequently, the leachate leaked to the outside of the coastal landfill sites is reduced by the low hydraulic conductivity as well as the sparser arrangement of joint sections in the SPSP cutoff wall, thus, significantly improving the leachate-containment effect.

The H-jointed SPSP cutoff wall with H-H joints (Case-IV) most efficiently achieves low hydraulic conductivity with a sparser arrangement of joint sections. The leakage of leachates in Case-IV can be traced to the lower reaches of the cutoff wall, occurring via the clay deposit layer, which is one of the bottom cutoff barriers and is further away than other pathways such as leakage directly through the cutoff wall (see Fig. 16). Thus, the H-jointed SPSP cutoff wall with the H-H joint sufficiently contributes to the leachate-containment effect of vertical cutoff barrier at coastal landfill sites.

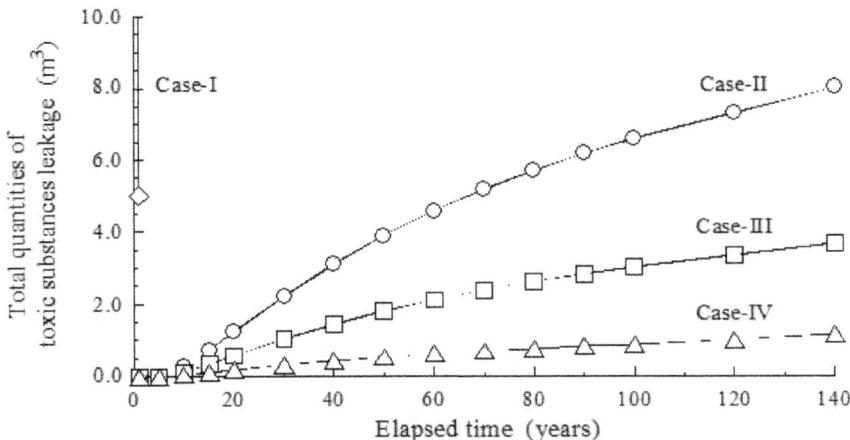

Figure 14: Total quantities of toxic substances leaked from SPSP cutoff wall onto sea side with respect to the elapsed time for Case-I to Case-IV

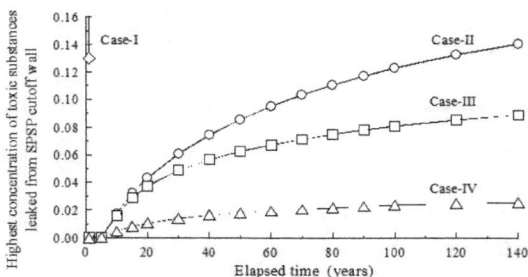

Figure 15: Relationships between elapsed time and the highest concentration of toxic substances leaked from SPSP cutoff wall onto sea side for Case-I to Case-IV

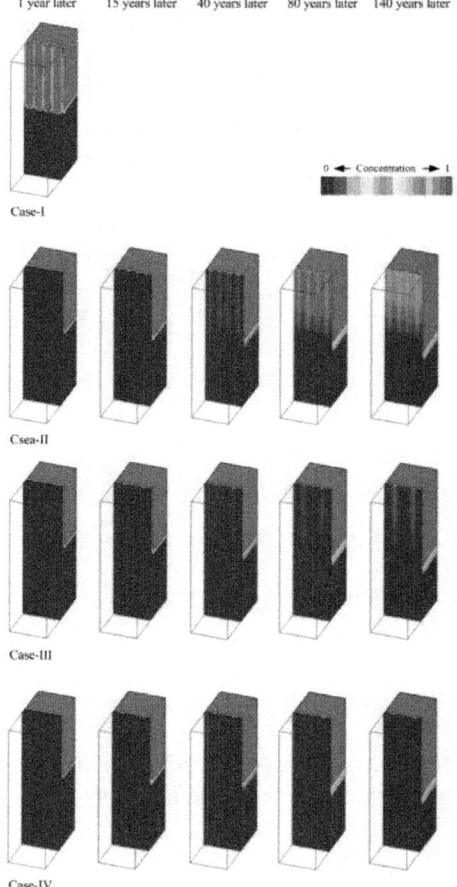

Figure 16: Distribution of concentration of toxic substances leaking out from waste layer in Case-I to Case-IV

In this section, it was clarified that technologies that lower the hydraulic conductivities of joint sections in SPSP cutoff walls and also facilitate the use of sparser arrangements contribute significantly to increasing the environmental feasibilities of SPSP cutoff walls at landfill sites. Further, the extent of the environmental feasibility of H-jointed SPSP cutoff walls with the H-H joints among the present technical developments in SPSP cutoff walls was shown.

CONCLUSIONS

An evaluation method that can express the local leakage of leachate from joint sections in steel pipe sheet pile (SPSP) cutoff walls is discussed, in this study. In particular, the evaluation of environmental feasibility (containment of leachates containing toxic substances) considering a three-dimensional arrangement and hydraulic conductivity distribution of the joint sections in the SPSP cutoff wall is compared with an evaluation that generally uses the equivalent hydraulic conductivity.

Evaluations of the environmental feasibilities of the SPSP cutoff walls with joint sections that have a higher hydraulic conductivity than that of the steel pipe must take into account the local leakage of leachates containing toxic substances from the joint section; this was possible using the SP/JS-model. Due to the local leakage into the surroundings of coastal landfills from joint sections, contamination in excess of the environmental standard values was confirmed to occur earlier than that predicted by the UL-model, which is the current standard evaluation method.

It was clarified that technologies that lower the hydraulic conductivities of joint sections in SPSP cutoff walls and also facilitate the use of sparser arrangements contribute significantly to increasing the environmental feasibilities of SPSP cutoff walls at landfill sites. Further, the extent of the environmental feasibility of H-jointed SPSP cutoff walls with the H-H joints among the present technical developments in SPSP cutoff walls was shown.

REFERENCES

1. S. Inazumi, M. Kimura, Y. Nishiyama, K. Yamamura, H. Tamura, M. Kamon, 2006New type of hydraulic cutoff walls in coastal landfill sites from H-jointed steel pipe sheet piles with H-H joints, Proceedings of 5th International Congress on Environmental Geotechnics, 1 725 732

2. Inazumi, S., Kimura, M., Too, A.J.K., & Kamon, M. (2005). Performance of H-jointed steel pipe sheet piles with H-H joint in vertical hydraulic cutoff walls, Proceedings of the 16th International Conference on Soil Mechanics and Geotechnical Engineering, 4 2269 2272

3.   Japanese Association for Steel Pipe Piles 1999 Steel Pipe Sheet Pile Foundations-Design and Construction- (in Japanese), Japanese Association for Steel Pipe Piles

4.   M. Kamon, T. Inui, 2002Geotechnical problems and solutions of controlled waste disposal sites (in Japanese), JSCE Journal of Geotechnical Engineering, 701/III-58, 1 15

5.   M. Kamon, Y. S. Jang, 2001Solution scenarios of geo-environmental problems, Proceedings of the 11th Asian Regional Conference on Soil Mechanics and Geotechnical Engineering, 833 852

6.   M. Kamon, T. Inui, T. Katsumi, M. Torisaki, 2007Risk control in the redevelopment of closed landfill sites from geotechnical viewpoint, Proceedings of the 7th Japanese-Korean-French-Seminar on Geo-Environmental Engineering, 19 24

7.   M. Kamon, T. Katsumi, K. Endo, K. Itoh, A. Doi, 2001Evaluation of the performance of coastal waste landfill with sheet pile containment system (in Japanese), Proceedings of the 5th Japan National Symposium on Environmental Geotechnology, 279 284

8.   M. Kimura, S. Inazumi, A. J. K. Too, K. Isobe, Y. Mitsuda, Y. Nishiyama, 2007 Development and application of H-joint steel pipe sheet piles in construction of foundations for structures Soils and Foundations 47 2 237 251

9.   M. Nishigaki, T. Hishiya, N. Hashimoto, I. Kohno, 1995The numerical method for saturated-unsaturated fluid density dependent groundwater flow with mass transport (in Japanese), JSCE Journal of Geotechnical Engineering, 501/III-30, 135 144

10. T. Oki, K. Torizaki, H. Kita, M. Yoshida, Y. Sakaguchi, H. Yoshino, 2003Evaluation of impermeability performance of the vertical impermeable walls by using steel sheet piles and steel pipe sheet piles (in Japanese), Proceedings of the 5th Japan National Symposium on Environmental Geotechnology, 53 58

11. K. Shimizu, 2003The latest geotechnical problems in waste landfill (in Japanese), Tsuchi-to-Kiso, Japanese Geotechnical Society, 51/58, 1 4

12. The Landfill System & Technologies Research Association of Japan 2004 Landfills in Japan (Rivised Edition) (in Japanese), The Journal of Waste Management

13. Waterfront Vitalization and Environment Research Center 2002 Design, Construction and Management Manual for Managed Type Waste Reclamation (in Japanese), Waterfront Vitalization and Environment Research Center

Chapter 3

EXPERIMENTAL STUDY OF DYNAMIC CHARACTERISTICS ON COMPOSITE FOUNDATION WITH CFG LONG PILE AND RAMMED CEMENT-SOIL SHORT PILE

Jihui Ding[1], Yanliang Cao[1], Weiyu Wang[2], Tuo Zhao[2], Junhui Feng[3]

[1]College of Civil Engineering, Hebei University, Baoding, China
[2]Hebei Academy of Building Research, Shijiazhuang, China
[3]China Metallurgical Design and Research Institute Co., Ltd., Baoding, China

ABSTRACT

Based on the idea of optimization design of pile type, the two kinds of the typical pile type are selected, which containing flexibility pile (e.g. rammed cement-soil pile is for short RCSP), and rigid pile (e.g. cement-flyash-gravel pile is for short CFGP). The three kinds of the composite foundation are designed, which are CFGP, CFG long pile and CFG short pile (for short CFGLP-CFGSP), CFG longshort pile and rammed cement-soil short pile (for short CFGLP-RCSSP). Natural earthquake is simulated by using the engineering blasting; the dynamic characteristics and dynamic response of the composite foundation are studied through field test. CFGLP-RCSSP is closed to linear relation. The bearing capacity of the four composite foundation of the CFGP, CFGLP-CFGSP, and CFGLPRCSSP in the site are 225 kPa, 179 kPa, and 197 kPa, separately increases 150%, 98.8% and 119% compared to the natural foundation. The vibration main frequency is mainly depended on properties of foundation soil and piles between vibration source and measuring point, pilling load value. Horizontal vibration main frequency greater than the vertical vibration main frequency and the vertical vibration main frequency close to the first-order natural frequency of composite foundation. With the pilling load increasing, the CFGLP-RCSSP pile composite foundation combined frequency decreased. Under the same blast energy, the acceleration peak on the CFG pile composite foundation is less than CFGLP-CFGSP the corresponding values, as the load increases, the peak acceleration gently. CFG pile composite foundation is favorable on seismic. The distribution of peak acceleration is consistent within 4 m from pile top in the CFGLP_RCSSP composite foundation. The maximum of the horizontal acceleration peak along the pile body occurs at a distance of pile top 4 m or the pile top, and that of vertical acceleration peak occurred at a pile top.

INTRODUCTION

The composite foundation is that the part soil body in the natural ground foundation is reinforced or replaced during the ground treatment, and load is born by reinforced body and soil body around the pile [1] . Design theory of single pile composite foundation is relatively mature, and has certain limitations and shortcomings. The pile stiffness is smaller and the pile body has certain bond strength in the flexible pile composite foundation. The most commonly flexible piles are mixing cement soil pile [2] [3] , rammed cement-soil pile [4] [5] pile, and so on. The strength of the flexible piles is low and load can not be effectively transmitted to the lower part of the pile. When the top of the soil-cement pile is crushed, the side friction of the pile length range did not develop out.

The pile in the rigid pile (i.e. CFG pile) composite foundation has higher strength, with large adjustment range of bearing capacity of composite foundation [6] . Usually the rigid pile has happened with piercing failure, and the pile body material strength has not fully developed out. To give full play to the advantages of various types of pile, the composite foundation to form by different typed piles combined together [7] [8] , can maximize the advantages of various types of pile. With the rapid development of economy, strength, length of pile in composite foundation can be greatly improved, and greatly improve the bearing capacity of composite foundation. The original design theory of composite foundation can not meet the requirements, and dynamic problems of composite foundation have become the focus of attention. Study on the seismic performance of the composite foundation is the main application of numerical analysis and simulation of the composite foundation of finite element software [9] -[11] , this method still remain at the theoretical level, have not been applied to the actual design. Wang Weiyu, Zhao Tuo, Ding Jihui etc. studied dynamic characteristic and its influence factors of cement soil pile and CFG pile composite foundation under the action of the blasting vibration [12] -[15] .

Optimization is made to CFG pile and rammed cement-soil pile, and the composite foundations of CFG pile, CFG long pile and CFG short pile,

CFG long pile and rammed cement-soil pile are, the designed. The stress and dynamic characteristics of the three composite foundation are studied through field tests.

THE INTRODUCTION OF THE TEST SITE

The test site is located in Shijiazhuang Heibei province. In the 20 m depth, the soil layers mainly are yellow silt clay, fine sand, middle sand and silt clay. In the 20 m driving depth, the underwater is not seen. There is not the harmful geologic action in the site. The main parameters of soil layer as shown in Table 1.

MODEL TEST AND SCHEME OF THE SITE

Three kinds of composite foundation model are designed: CFGP, CFGLP-CFGSP, CFGCP-RCSLP composite foundation. The model design parameter of composite foundation is shown in Table 2. CFG pile adopts C20 commercial concrete. Blasting is used as the vibration resource. The diameter of blasting hole is 50 mm. The Explosives are buried in the hole and than backfill tamping. The vibration is picked by acceleration sensors. The arrangement of the piles and measuring elements are shown in Figures 1-4. The upper load is supplied by pilling concrete block, and each load of the composite foundation is added by an electric pressure pump-hydraulic jack. The square steel is 2.0×2.0 m as the loading plate.

THE ANALYSIS OF EXPERIMENTS RESULT

Load-Settlement Curves

Combining the three kinds load test of composite foundation, the load-settlement curves as shown in Figure 4. From Figure 4, compared with natural foundation, the bearing capacity of composite foundation of CFGP, CFGLP-CFGSP and CFGLP-RCSSP increases obviously and the deformation of composite foundation de-

Table 1: Mainly parameter of soil layer

No.	h_i/m	f_{ak}/kpa	E_s/Mpa	f_s/kpa	f_{pk}/kpa
1)	0.5	130	6.17	60	
2)	1.5	140	10	55	800
3)	4.0	200	11.5	82	1500
4)		270	5.84	65	1000

Where, h_i is thickness of the soil layer, f_{ak} is characteristic value of bearing capacity; E_s is compression Modulus. f_{sk} is ultimate shaft resistance, f_{pk} is ultimate tip resistance.

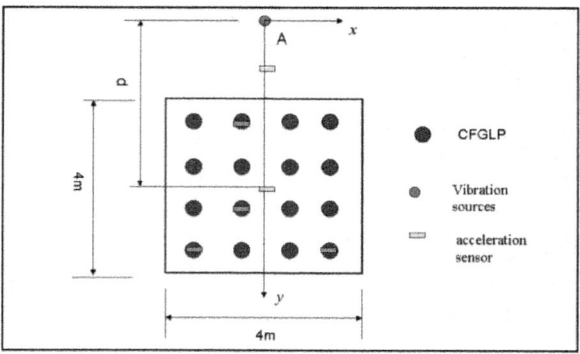

Figure 1: The arrangement of the piles and measuring elements of the Model 1.

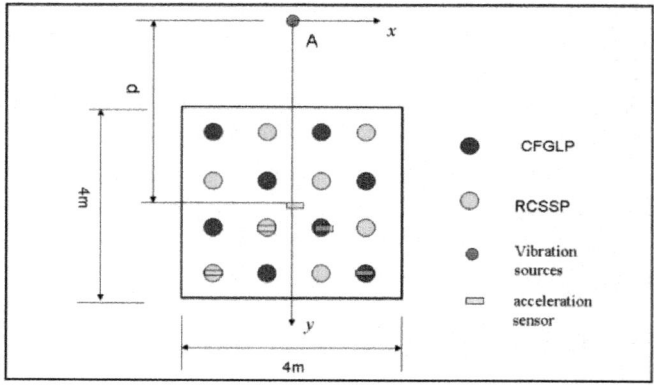

Figure 2: The arrangement of the piles and measuring elements of the Model 2.

creased compared to the Natural Foundation. The nonlinear degree of the p - s curves of combined pile decreases, and CFGLP-RCSSP is closed to linear relation. The bearing capacity of the four composite pile of the CFGP, CFGLP-CFGSP, and CFGLP-RCSSP in the site are 225 kPa, 179 kPa, and 197 kPa, separately increases 150%, 98.8% and 119% compared to the Natural Foundation.

Table 2: Model design parameter of composite foundation

Model	Type	Pile length/m	Pile diameter/mm	Pile spacing/mm	Replacement rate
1	CFGP	6.0	350	100	0.09616
2	CFGLP	6.0	350	200	0.04808
	CFGSP	4.0	350	200	0.04808
3	CFGLP	6.0	350	200	0.04808
	RCSSP	4.0	350	200	0.04808

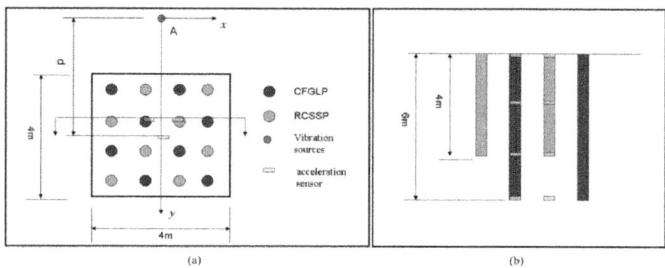

Figure 3: The arrangement of the piles and measuring elements of the Model 3. (a) Plane arrangement;(b)1-1profilearrangement.

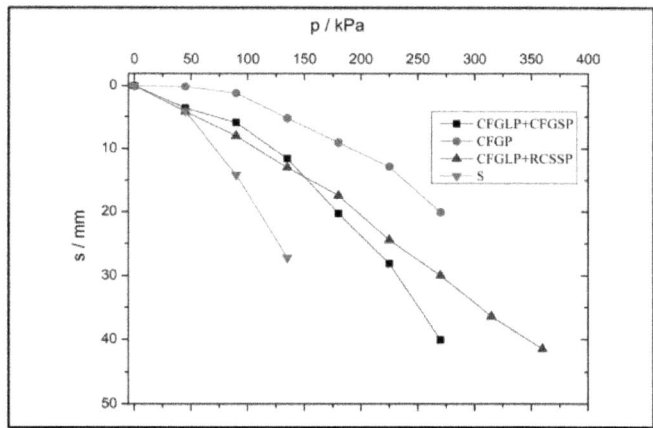

Figure 4. The p - s curves of the composite foundation.

The Main Frequency of Vibration

Field test shows that, under the same blast energy, vibration position, properties of foundation soil and form of composite foundation, have influence on the main frequency of composite foundation, but the rule is not obvious. Vibration main frequency depends on the properties of the soil and pile body between the vibration source and measuring point. When no load, the horizontal vibration main frequency of the natural foundation is at 8.2 - 9.03 Hz, and the vertical vibration main frequency is at 7.75 - 9.16 Hz; the horizontal vibration main frequency CFGP composite foundation is at 21.1 - 27.7 Hz, and the vertical vibration frequency is at 6.8 - 31.6 Hz; the horizontal vibration main frequency of the CFGLP-CFGSP composite foundation is at 17.4 - 29.6 Hz, and the vertical vibration main frequency is at 7.2 - 9.4 Hz; the horizontal vibration main frequency of the CFGLP-RCSSP composite foundation is at 7.6 - 38.4

Hz, and the vertical vibration main frequency is at 8.2 - 43.1 Hz. Horizontal vibration main frequency greater than the vertical vibration main frequency, and the vertical main vibration frequency close to the natural frequency of composite foundation. With the load increasing, the main vibration frequency of the CFGLP-RCSSP composite foundation decreases. When the load is 180 kPa, the horizontal vibration main frequency of the CFGLP-RCSSP composite foundation is at 7.4 - 22.6 Hz, and the vertical vibration main frequency is at 7.3 - 28.0 Hz.

Peak Acceleration Results

Figures 5-7 are the peak acceleration with the horizontal distance r from the measuring point to the vibration source without no pilling load on the CFGP composite foundation, when the depth of the vibration source is 6 m and the explosive quantity is 1.05 kg. From Figure 4, in addition to individual point the horizontal peak acceleration is greater than the vertical peak acceleration, and with the increase of r, the difference gradually decreases. From Figures 5-7, outside the scope of the composite foundation, the peak acceleration significantly decreased with the increase of the pilling load; in the surface of compound foundation, pilling load action makes the peak acceleration gently.

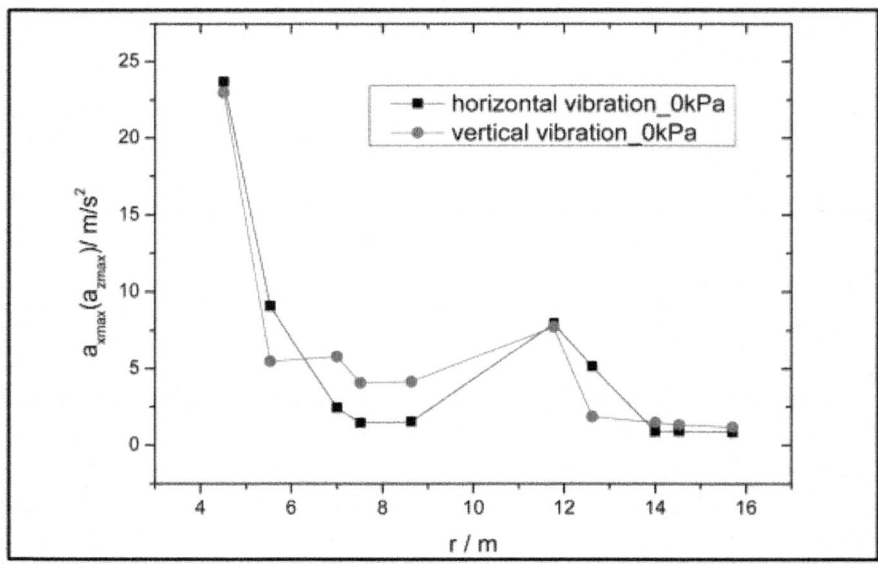

Figure 5: $a_{xmax}(a_{zmax})$-r of CFGP composite foundation (no load).

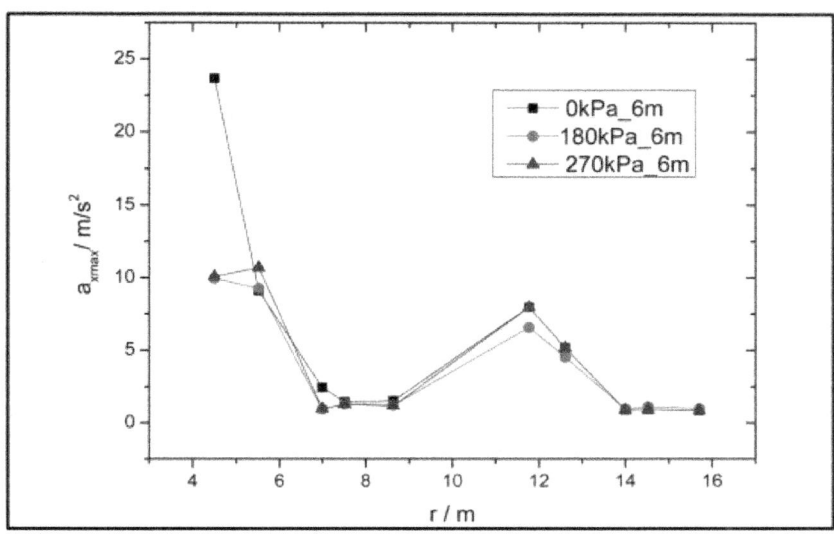

Figure 6: a_{xmax}-r of CFGP composite foundation.

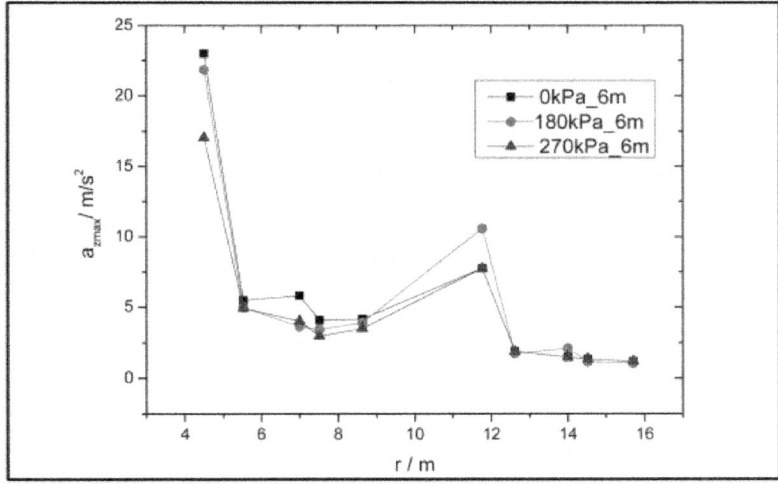

Figure 7: a_{zmax}-r of CFGP composite foundation.

When the depth of the vibration source is 6 m and r = 7 - 8.6 m, with the no-load, the horizontal acceleration peak of the CFG pile composite foundation surface is at 2.44 - 1.53, and the vertical acceleration peak is 5.8 - 4.2; while pilling load is 270 kPa, the horizontal acceleration peak is 1.00 - 1.17, and vertical acceleration peak is 4.01 - 3.96.

When the distance from vibration source to the center of CFGP composite foundation distance is 14 m and the depth of vibration source is 6 m from the ground, the ratio of horizontal acceleration and vertical acceleration peak is at 0.61 - 2.75, the measuring point outside CFGP composite foundation when the distance r is 1 m, the ratio was 1.03, and near to 1.0. When the distance from vibration source to the center of CFGP composite foundation distance is 7 m and the depth of vibration source is 6 m from the ground, the ratio of horizontal acceleration and vertical acceleration peak is at 0.36 - 1.65; the farther the horizontal distance r from the measuring point to vibration source is, the greater the ratio.

Figures 8 and 9 are the peak acceleration with the horizontal distance r from the measuring point to the vibration source without no pilling load on the CFGLP-CFGSP composite foundation, when the depth of the vibration source is 6 m and the explosive quantity is 1.05 kg. From Figures 8-10, the peak acceleration along the CFGLP on the CFGLP-CFGSP composite foundation is near to peak acceleration of the CFGSP.

When the depth of the vibration source is 6 m and r is at 7 - 8.6 m, with the no-load, the horizontal peak acceleration of the CFGLP-CFGSP composite foundation surface is at 9.05 - 1.53 m/s^2, and the vertical peak acceleration is 13.27 - 2.72 m/s^2; while r is at 14 - 15.7 m, the horizontal peak acceleration is 5.23 - 2.25 m/s^2, and vertical peak acceleration is 5.13 - 0.7 m/s^2.

When the distance from vibration source to the center of CFGLP-CFGSP composite foundation distance is 14 m and the depth of vibration source is 6 m from the ground, the ratio of horizontal acceleration and vertical acceleration peak is at 1.02 - 3.15, measuring point distance from the vibration source is equal, the ratio is 1.49 on the CFGSP measuring point, and ratio is 3.15 on the CFGLP measuring point.

When the distance from vibration source to the center of CFGLP-CFGSP composite foundation distance is 7 m and the depth of vibration source is 6m from the ground, the ratio of horizontal acceleration and vertical acceleration peak is at 0.69 - 1.91, measuring point distance from the vibration source is equal, the ratio is 1.91 on the CFGSP measuring point, and ratio is 1.71 on the CFGLP measuring point. the farther the horizontal distance r from the measuring point to vibration source is, the greater the ratio.

Figures 11-18 are CFGLP and RCSSP peak acceleration changes of the CFGLP-RCSSP composite foundation with vibration source depth and pilling load, when the explosive quantity is 1.05 kg and pilling load is 360 kPa.

Figure 11 is the distribution law of the horizontal vibration peak acceleration on the CFGLP of the CFGLPRCSSP composite foundation. The depth of the vibration source is separately 2.5 m and 6 m, the horizontal vibration peak acceleration distribution is almost consistent. When the depth location of the vibration source is 7 m, the horizontal vibration peak acceleration maximum is at z = 4 m; when the location of the vibration source is 14 m, the horizontal vibration peak acceleration maximum is at z = 0 m.

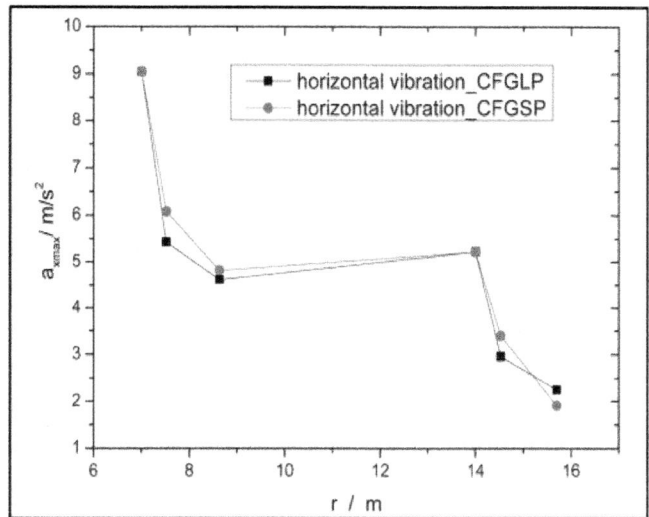

Figure 8: a_{xmax}-r of CFGLP-CFGSP composite foundation.

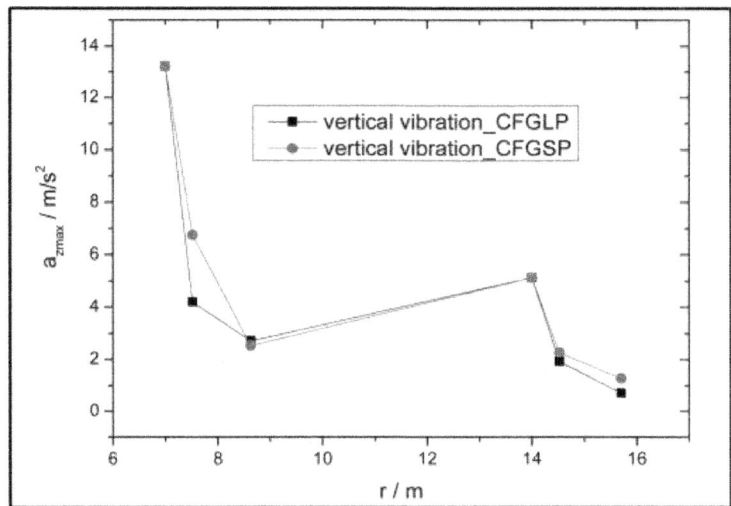

Figure 9: a_{zmax}-r of CFGLP-CFGSP composite foundation.

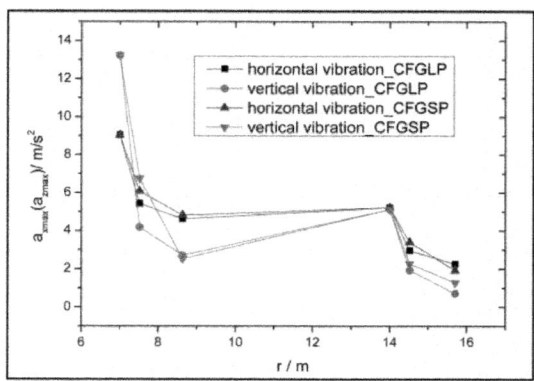

Figure 10: $a_{xmax}(a_{zmax})$-r of CFGLP-CFGSP composite foundation.

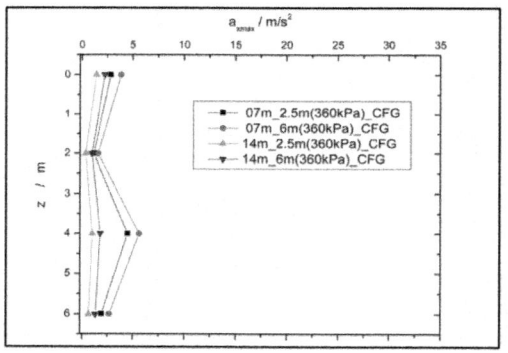

Figure 11: The peak acceleration a_{xmax} along the CFGLP.

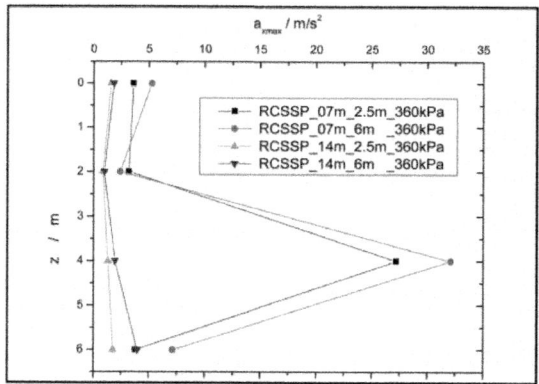

Figure 12: The peak acceleration a_{xmax} along the RCSSP.

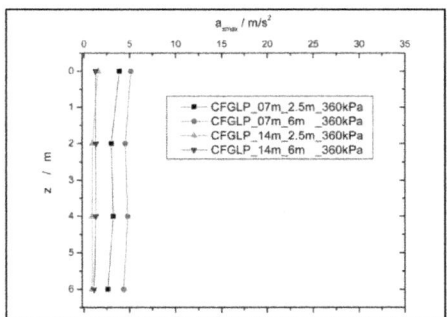

Figure 13: The peak acceleration a_{zmax} along the CFGLP.

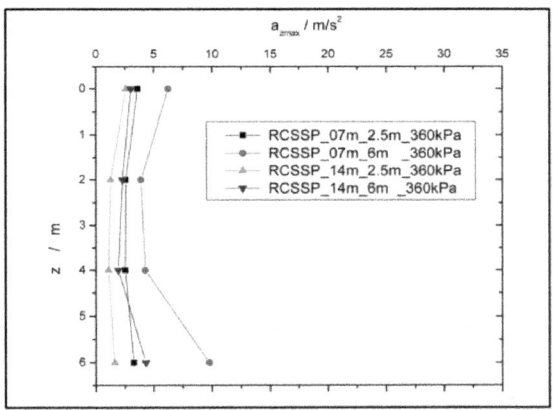

Figure 14: The peak acceleration a_{zmax} along the RCSSP.

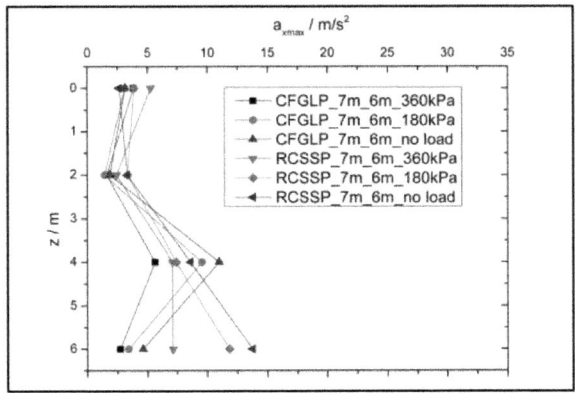

Figure 15: a_{xmax}-z in CFGLP-RCSSP composite foundation.

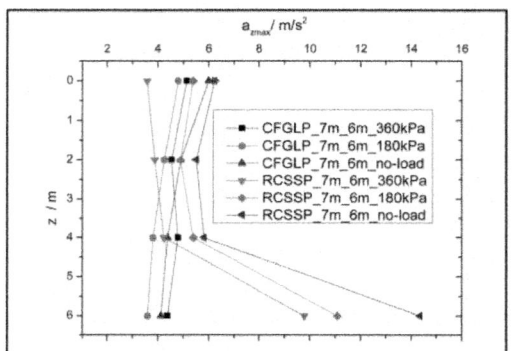

Figure 16: a_{zmax}-z in CFGLP-RCSSP composite foundation.

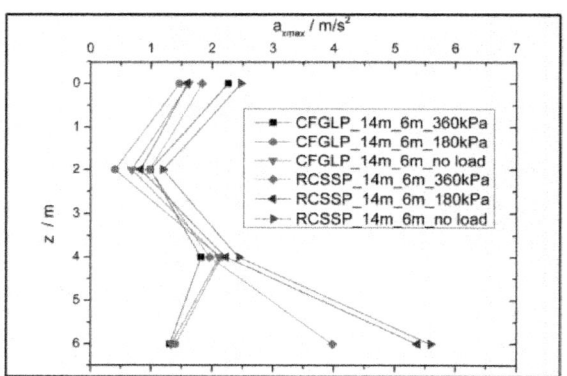

Figure 17: a_{xmax}-z in CFGLP-RCSSP composite foundation.

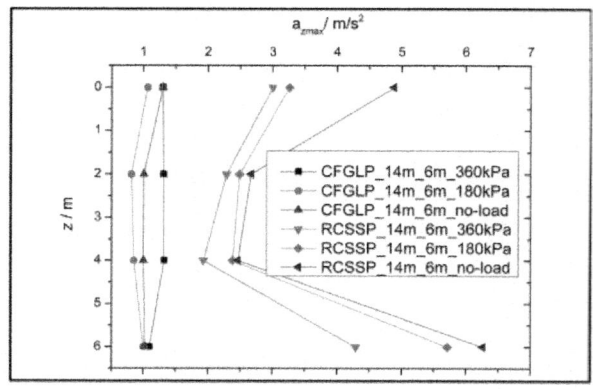

Figure 18: a_{zmax}-z in CFGLP-RCSSP composite foundation.

Figure 12 is the distribution law of the horizontal vibration peak acceleration on the RCSSP of the CFGLPRCSSP composite foundation. When the depth location of the vibration source is 7 m, the horizontal vibration peak acceleration maximum is at $z = 4$ m; when the location of the vibration source is 14 m, the horizontal vibration peak acceleration maximum is at $z = 6$ m.

Figure 13 is the distribution law of the Vertical vibration peak acceleration on the CFGLP of the CFGLPRCSSP composite foundation. The vertical vibration acceleration peak on the CFGLP is change little, and the vertical vibration peak acceleration maximum is at $z = 0$ m.

Figure 14 is the distribution law of the Vertical vibration peak acceleration on the RCSSP of the CFGLPRCSSP composite foundation. The vertical vibration acceleration peak on the RCSSP is change little, and the vertical vibration peak acceleration maximum is at $z = 0$ m.

Figures 14-18 are the distribution law of the vibration peak acceleration along the pile of the CFGLP-RCSSP composite foundation. With the increase of pilling load on CFGLP-RCSSP of the composite foundation the peak acceleration is reduced, the change range of composite foundation of 4 m pile body acceleration is consistent, the peak acceleration of the measuring point of the RCSSP is larger than that of the CFGLP. The maximum of the horizontal acceleration peak occurs in $z = 4$ m or $z = 0$ m. The maximum of the vertical acceleration peak occurs in $z = 0$ m.

CONCLUSIONS

The nonlinear degree of the p - s curves of combined pile composite foundation decreases, and CFGLP-RCSSP is closed to linear relation. The bearing capacity of the four composite piles of the CFGP, CFGLP-CFGSP, and CFGLP-RCSSP in the site are separately 225 kPa, 179 kPa, and 197 kPa, separately increases 150%, 98.8% and 119% compared to the natural Foundation.

The field test shows that, under the same blast energy, vibration source position, form of composite foundation and properties of foundation soil, influence the main frequency of the composite foundation, but the rule is not obvious. The vibration main frequency is mainly depended on properties of foundation soil and piles between vibration source and measuring point, pilling load value. Horizontal vibration main frequency greater than the vertical vibration main frequency and the vertical vibration main frequency close to the first-order natural frequency of composite foundation. With the pilling load increasing, the CFGLP-RCSSP pile composite foundation combined frequency decreased. The field test shows that, under the same blast energy, vibration source position, the acceleration peak on the CFGP composite foundation is less than CFGLP-CFGSP the corresponding values, as the load increases, the peak acceleration gently. CFGP composite foundation is favorable on seismic.

Field test shows that, under the same blast energy, vibration source positions, form of composite foundation, properties of foundation soil and pilling load have a significant effect on the peak acceleration of composite foundation. The distribution of peak acceleration is consistent within 4 m from pile top in the CFGLP-RCSSP composite foundation. The maximum of the horizontal acceleration peak along the pile body occurs at a distance of pile top 4 m or the pile top, and that of vertical acceleration peak occurred at a pile top.

REFERENCES

1.   (2002) GB50007-2002 Code for Design of Building Foundation(s). China Architecture Industry Press, Beijing.

2.   Yoshio, S. (1983) Deep Mixing Chemical Method Using Cement as Hardening Agent. Symposium on Soil and Rock Improvement Techniques, Bangkok, 34-37.

3.   Horii, N., Toyosawa, Y., Tamate, S. and Hashizume, H. (1998) Stability of Composite Ground Improved by Deep Mixing Method. Proceedings of 2nd International Conference on Ground Improvement Techniques, Singapore, 193- 198.

4.   Zheng, G. and Jiang, X.L. (1999) Research on the Bearing Capacity of Cement Treated Composite Foundation. Rock and Soil Mechanics, 3, 46-50.

5.   Ma, H.L. (2003) Quantitative Analyses of the Influence of Pile Length and Other Factors on Capability and Modul of Cement Stabilized Soil Composite Foundation. Chinese Journal of Geotechnical Engineering, 11, 720-723.

6.   Fu, J.H. and Song, E.X. (2000) Analysis of Rigid Pile Composite Foundation's Working Performance. Rock and Soil Mechanics, 21, 335-339.

7.   Wang, M.-S., Wang, G.-C., Yan, X.-F., et al. (2005) In-Situ Tests on Bearing Behavior of Multi-Type-Pile Composite Subgrade. Chinese Journal of Geotechnical Engineering, 27, 1142-1145.

8.   Yan, M.L., Wang, M.S., Yan, X.F. and Zhang, D.G. (2003) Study on the Calculation Method of Multi-Type-Pile Composite Foundation. Chinese Journal of Geotechnical Engineering, 25, 352-355.

9.   Ding, J.H. (2007) Reliability Analysis on the Bearing Capacity of Composite Foundation with Multi-Type Compound Piles. Engineering

Mechanics, 25, 168-172.

10. Ding, J.H., Liu, F.R. and Du, E.X. (2008) Dynamic Characteristic Analysis on Composite Foundation with Soil-Cement Piles and CFG Piles. Fly-Ash Comprehensive Utilization, 6, 37-40.

11. Wang, W.Y., Zhao, T. and Meng, Y.J. (2012) The Numerical Analysis on Dynamic Characteristics of CFG Pile Composite Foundation under Blasting. Engineering Mechanics, 29, 150-155.

12. Wang, W.Y., Zhao, T. and Ding, J.H. (2011) Effects on Dynamic Characteristics and Response of Rammed Soil-Cement Pile Composite Foundation. Engineering Mechanics, 28, 187-191.

13. Wang, W.Y., Zhao, T. and Ding, J.H. (2010) Influence Factors of Dynamic Characteristics and Response of CFG Pile Composite Foundation. Chinese Journal of Geotechnical Engineering, S2, 115-118.

14. Zhao, T., Yang, C.M. and Wang, W.Y. (2010) The Dynamic Test Research on CFG Pile Composite Foundation under Blasting. Journal of Highway and Transportation Research and Development (Applied Technique), 7, 121-122.

15. Ding, J.H., Wang, W.Y., Zhao, T., et al. (2013) The Dynamic Characteristic Experimental Method on the Composite Foundation with Rigid-Flexible Compound Piles. Open Journal of Civil Engineering, 3, 94-98. http://dx.doi.org/10.4236/ojce.2013.32010

Chapter 4

NEW METHOD FOR PREDICTION PILE CAPACITY EXECUTED BY CONTINUOUS FLIGHT AUGER (CFA)

Wael N. Abd Elsamee

Faculty of Engineering, Sinai University, El Arish, Egypt

ABSTRACT

A study of piles is quit complex and the estimation of carrying capacity is calculated from theoretical formula and load test results. The design resistance may be calculated using conventional static pile design theory. The pile founding depths should be predetermined before installation from a site geotechnical investigation. To ascertain the field performance and estimate load carrying capacities of piles, in-situ pile load tests should be conducted. In this study, field pile load test data is analyzed to estimate the ultimate load for end bearing piles. The investigated site is about 100×110 m located in Alexandria, Egypt. Geotechnical investigations at the site are carried out to a maximum depth of 45 m. Four borings have been done in field. The tests are conducted at the site for two skelton structure buildings to be constructed on raft foundation rested on piles executed by continuous flight auger. Four pile load tests are performed on 600 mm diameters and 27 m lengths. Ultimate capacities of piles are determined according to different methods. It is concluded that the percentage of friction load carried by the shaft along the pile length is about 46% of total load while the percentage of load carried by the end bearing is 54% of total load. A new proposed method by the author is presented to calculate the ultimate capacity of pile from pile load test. The proposed method depends on the settlement of pile without taken into consideration the elastic deformation. An empirical formula is presented from the relationship between stress and settlement of pile due to friction and end bearing only after deducting the elastic deformation. However, the obtained results for the ultimate capacity of end bearing piles are considered to be more accurate than other methods. The proposed method appears to give bitter results that agrees well with the theoretical predictions. The proposed method is easier, quicker and more reliable.

INTRODUCTION

Piles are relatively long and generally slender structural foundation members that transmit superstructure loads to deep soil layers. Today, there are numerous types of piles being developed and extensively use in the construction industry. However, difference pile system will serve difference purposes in different type of soil and site conditions. Generally most of the piles are design to meet the requirements of the end bearing capacity which is driven to set on to the hard strata. However, pile section also can generate certain percentage of resistance through skin friction that produced between the pile and soil.

The prediction of the axial capacity of piles has been a challenge since the beginning of the geotechnical engineering profession.

Ir, T. Y. C., Chow, C. M. G. and Partners, S. B. (2003) presented some aspect of design and construction of bored pile foundation in Malaysia. Empirical equations correlating the value of the ultimate shaft resistance (f_{su}) and the ultimate base resistance (f_{bu}) to SPT'N' values are suggested as design of bored piles under axial compression load. Some aspects of design and construction in difficult ground conditions such as limestone and soft ground were presented together with some suggestions on quality control for bored pile construction [1].

Dan, A. B., Steven, D. D., Robert, W. T. and Carlos, A.L. (2007) introduced a manual of the state-of-thepractice for design and construction of continuous flight auger (CFA) piles, including those piles commonly referred to as augured cast-in-place (ACIP) piles. Quality control (QC)/quality assurance (QA) procedures were discussed, and general requirements for a performance specification are given. Methods to estimate the static axial capacity of single piles were recommended based on a thorough evaluation and comparison of various methods used in the United States and Europe. A generalized step-by-step method for selecting and designing CFA piles was presented [2].

Akbar, A., Khilji, S, Khan, S.B, Qureshi, M.S. and Sattar, M (2008) presented the experience gained from four pile load tests at a site in the North West Frontier Province of Pakistan. Geotechnical investigations at the site were carried out to a maximum depth of 60 m. The sub soils at the site are predominantly hard clays within the investigated depth with thin layers of gravels/ boulders below 40 m depth. Four piles of diameter varying from 660 mm to 760 mm and length ranging between 20 m and 47.5 m were subjected to axial loads. Using the pile load test results, back calculations were also carried out to estimate the appropriate values of pile design parameters [3]. Kenji, I. (2010) presented a brief introduction of the in-situ pile loading tests that have been conducted in Japan over the last two decades in connection with the design and construction of high-rise buildings in areas of soft soil deposits. In addition to the conventional types of tests in which the load is applied at the top and at the toe of the pile (O-cell test), what may be called "pile toe bearing test" and "skin friction test" was introduced. The results of these tests were described and compared with those from the conventional type of the pile loading tests. In-situ prototype tests are also introduced in which bearing power of Barrette type pile was compared with that of the circular type pile. A special case of in-situ pile loading tests conducted in Singapore was also introduced in which the friction between the circular ring-shaped concrete segment and the surrounding soil deposit was measured directly during excavation of the shaft by applying loads up and down by jacks installed between two adjacent segments in vertical direction. The whole scheme and process of construction is for these two undertakings were introduced with some comments on observed behaviour of the walls and on special precaution taken during construction [4].

Manandhar, S. and Yasufuku, N. (2011) presented the mechanism of tapered pile through small scale model tests. The increment of effective failure zone around the pile tip area with increasing tapering angle was discussed. On the load-settlement curve during pile penetrate, evidences of model tests showed the increase in end bearing behavior by tapered piles. The analytical spherical cavity expansion theory had been utilized to evaluate the end bearing capacity. In the proposed model, the effects of angle of tapering have been introduced to compute the end bearing capacity of tapered piles. The test results and the proposed model showed that the tapering angle affects the end bearing resistance comparing with conventional straight piles on different types of sands at different relative densities. The studies incorporating model tests, prototype tests and real type pile tests have been validated and predicted well the proposed model [5].

Wael N. Abd Elsamee (2012) presented field pile load tests data which was analyzed to estimate the ultimate load for friction piles. The analysis was based on three pile load test results. The tests were done at the site of The Cultural and Recreational Complex Project in Port Said-Egypt. Three pile load tests were performed on bored piles of 900 mm diameter and 50 m length. Geotechnical investigations at the site were carried out to a maximum depth of 60 m. Ultimate capacities of piles were determined according to different methods. It was concluded that about 8% of load is resisted by the pile at the base, and that up to 92% of load is resisted by friction shaft. A new proposed method to calculate ultimate capacity of pile from pile load test was presented

[6]. From the above, the variation in the load estimates of available methods is big. Thus, additional study on End Bearing pile capacity is needed to be done. However, the objective of this study is to analysis the results of actual pile tests and to develop a formula for closer prediction of the pile capacity.

SOIL INVESTIGATION

The investigated site is about 100×110 m located in Alexandria, Egypt. Geotechnical investigations at the site were carried out to a maximum depth of 45 m. Four borings have been done in field for investigations. Figure 1 shows the soil profile of the investigated site. However, the following soil stratifications were encountered:

1) From elevation (0.00) to (−2.00) Silty clay with percentage of broken shells (Fill).

2) From elevation (−2.00) to (−10.00) Soft silty clay with percentage of sand and traces of broken shells.

3) From elevation (−10.00) to (−13.50) Soft silty clay with percentage of shells.

4) From elevation (−13.50) to (−16.50) Soft silty clay.

5) From elevation (−16.50) to (−18.00) Fine to medium sand.

6) From elevation (−18.00) to (−20.00) Hard silty clay.

7) From elevation (−20.00) to (−21.50) Graded sand.

8) From elevation (−21.50) to (−23.50) Hard silty clay.

9) From elevation (−23.50) to (−45.00) Graded sand.

The ground water table has been found to be at 1.25 meter from the ground surface. According to the geotechnical investigations pile foundation is recommended. The pile founding depths should be predetermined before installation. The lengths of piles were to be taken 27 m with 600 mm diameter according to the soil investigation as well as the theoretical prediction.

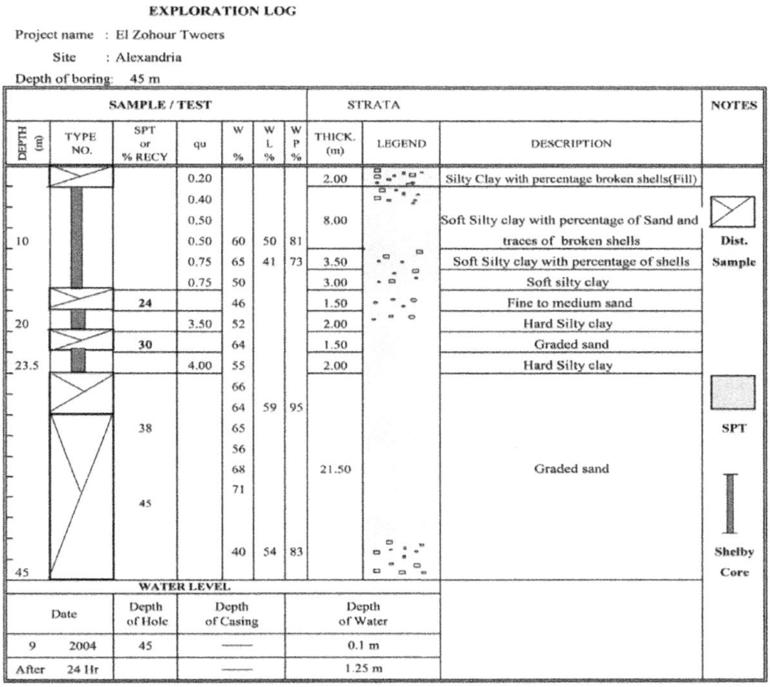

EXPLORATION LOG

Project name : El Zohour Twoers

Site   : Alexandria

Depth of boring: 45 m

DEPTH (m)	TYPE NO.	SPT or % RECY	qu	W %	W L %	W P %	THICK. (m)	LEGEND	DESCRIPTION	NOTES
			0.20				2.00		Silty Clay with percentage broken shells(Fill)	
			0.40							
			0.50				8.00		Soft Silty clay with percentage of Sand and	
10			0.50	60	50	81			traces of broken shells	Dist.
			0.75	65	41	73	3.50		Soft Silty clay with percentage of shells	Sample
			0.75	50			3.00		Soft silty clay	
		24		46			1.50		Fine to medium sand	
20			3.50	52			2.00		Hard Silty clay	
		30		64			1.50		Graded sand	
23.5			4.00	55			2.00		Hard Silty clay	
				66						
				64	59	95				
		38		65						SPT
				56						
				68			21.50		Graded sand	
				71						
		45								
				40	54	83				Shelby
45										Core

WATER LEVEL

Date		Depth of Hole	Depth of Casing	Depth of Water
9	2004	45	———	0.1 m
After	24 Hr		———	1.25 m

Figure 1: Soil profile of the investigated site.

3. Theoretical Prediction of Pile Load Capacity

A study of piles is quit complex and the estimation of carrying capacity is calculated from theoretical formula and load test results. Before execution of piles, estimation of pile load capacity is done by theoretical formula as follows:

$$Q_u = Q_s + Q_b \qquad (1)$$

where:

Q_u = ultimate pile capacity;

Q_s = ultimate shaft resistance = surface area of shaft in contact with the soil $\times$ shear strength of the soil;

Q_b = ultimate base resistance;

$$Q_b = q_b \times A_b \qquad (1.2)$$

$$Q_s = C \times d \times L \, (\text{clays})$$
$$Q_s = f_s \times d \times L \, (\text{sands})$$

(1.3)

f_s = skin friction;

d = diameter of pile;

L = length of pile in contact with the soil;

C = adhesion;

q_b = base bearing capacity;

A_b = area of base.

Figure 2 shows the vertical loads (shaft resistance and base resistance) of pile.

The design resistance may be calculated using conventional static pile design theory. The theoretical pile capacities have been calculated by using Egyptian code (2005). The following Equations (2) and (3) are used to calculate the ultimate pile capacity of end bearing pile [7]. (Pile diameter used = 600 mm and Pile length = 27 m).

$$Q_{all} = 90N \left(\pi R^2 \right) + N' \left(2\pi RL \right)$$

(2)

where:

Q_{all} = the working pile load at $F \cdot S = 2.5 (\text{kN})$ for end bearing and $F \cdot S = 2.0 (\text{kN})$ for shaft friction;

N = the average value of number of blows in (S.P.T) test for the effective soil at end bearing from distance (2R) blow base of pile;

N' = the average number of blows in (S.P.T) test

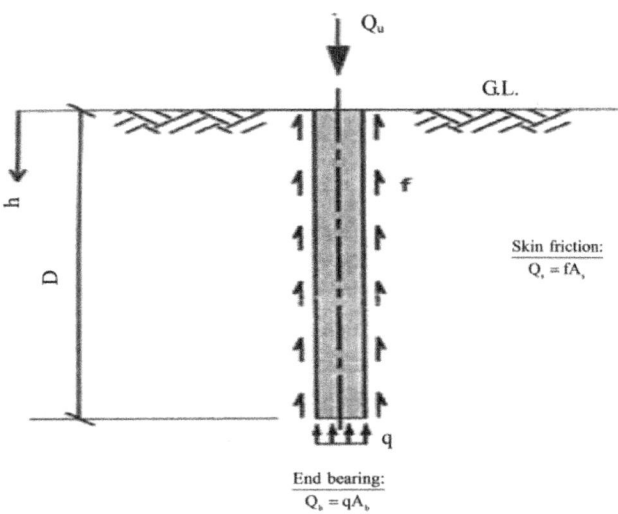

Figure 2: Vertical loads (shaft resistance and base resistance) of pile.

along of the pile length inside the layers of cohessionless soil;

R = radius of pile (meter) = 0.30 m;

L = length of penetration pile layers of cohessionless soil (meter)) = 7.50 m;

At investigated site the ultimate capacity of pile by using Equation (2) is $Q_u = 3733.64\text{kN}$.

However, Egyptian code (2005) state that the ultimate capacity of pile can be estimated as follows:

$$Q_u = fA_s + A_b\left(q - p_o\right) \tag{3}$$

where:

Q_u = ultimate capacity of pile;

A_s = side area of pile length;

A_b = base bearing area;

f = average stress (friction or adhesion);

q = average pressure stress on cross section of pile at base of pile;

$p_o =$ own weight of pile.

Table 1 shows the calculated theoretical ultimate pile capacity for pile diameter 60 cm with length 27.0 m. From this table it can be shown that the obtained Q_{u1t} = 2250.08 Kn/m^2 and after taken factor of safety (F·S = 2.0) then the $Q_{all} = 1125.04\,\mathrm{kN/m^2}$. However, the allowable bearing capacity of pile for the investigated site is taken 1000 = kN/m^2. **Figure 3** shows the calculated ultimate capacity of the pile using Equations (2) and (3).

However, it can be shown from Equation (3) and **Figure 3** that the percentage of friction load carried by the shaft along the pile length is about 46% of total load, while the pile load carried by the end bearing is 54% of total load.

CONSTRUCTION METHOD OF PILES BY CONTINUOUS FLIGHT AUGER (CFA)

The drilling process for (CFA) is suitable for penetrating dense layers and is unaffected by ground water or collapsing soil conditions. However, CFA method can be summarized as follows:

- The pile is formed by first drilling into the ground with a continuous flight auger to the required depth;
- Concrete is then injected under pressure through the auger's hollow stem as it is being withdrawn;
- The concrete pressure is maintained during the auger withdrawal so that it assists the extraction as well as exerting a lateral pressure on the surrounding soils;
- Reinforcing cage is placed into the concrete column. **Figure 4** shows the steps of execution process by con-tinuous flight auger (CFA) piles. Figures 5-8 show the execution process by continuous flight auger (CFA) piles at the site.

Table 1: The calculated theoretical ultimate pile capacity for pile diameter 60cm with length 27.0 m

Layer	Layer Depth	Soil Type	Av. SPT N Value	Undrainage cohesion		Values according ECP (2004)							
	Under the SBL [m]			Cu [kN/m²]	SPT	Pile own weight	Pile area m²	Layer Thicknes	Skin friction τ [kN/m²]	Bearing stress [KN/m²]	Pile Perimeter [m]	Skin friction [kN/m]	Friction Force Q [KN]
1)	0.0 - 16.5	SS-c		5			0.28	16.5	2.5		1.884955	77.8	77.8
2)	16.5 - 18.0	MS-S	24		20 - 30		0.28	1.5	2.86		1.884955	8.1	85.8
3)	18.0 - 20.0	HS-C		35			0.28	2	17.5		1.884955	66.0	151.8
4)	20.0 - 21.5	GS-S	30		20 - 30		0.28	1.5	3.02		1.884955	8.5	160.4
5)	21.5 - 23.5	HS-C		40	-		0.28	2	20		1.884955	75.4	235.8
6)	23.5 - 28.0	GS-S	38		38 - 45		0.28	4.5	9.43		1.884955	80.0	315.7
8)	base of pile	GS-S		4515.73		190.85	0.28			1222.831			
								Qt =		1222.83			1027.25
										Qu =	2250.08		

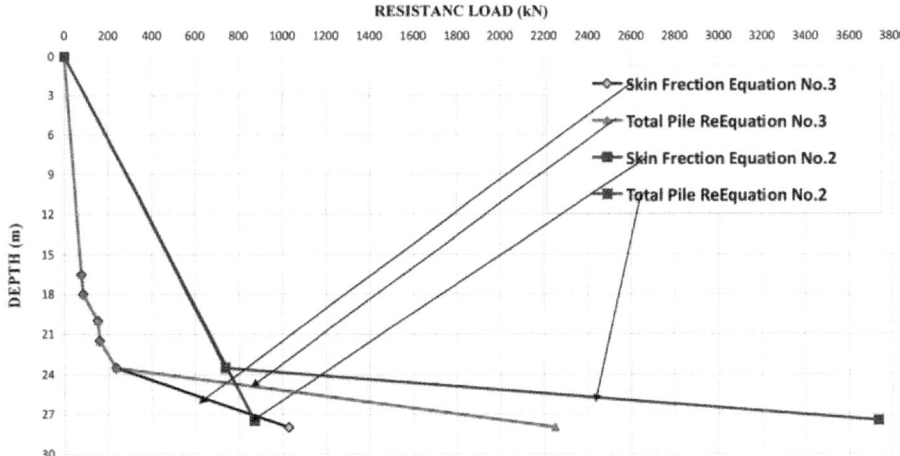

Figure 3: The relationship between the calculated theoretical capacity and depth of pile.

PILE LOAD TESTS

Load tests are performed to proof the design load and to check the pre-chosen factor of safety. In this study four pile load tests were performed on 600 mm diameter and 27 m lengths. The tests were done according to Egyptian Code. The test pile cap is shown in Figure 9. The reaction load was performed by a system of jacking bearing against dead load resting on a platform (Kentledge). The dead load was supplied by plastic sand bags on the platform as shown in Figure

10. The platform was supported on three wide flanged girder beams (reaction beams) placed side by side (and bolted together) over the jack. A hydraulic jack system comprising a 550 tons jack, pressure gauge, oil reservoir, pump and piping was used in the test. Settlement of the pile was recorded by means of four settlement dial gauges capable of reading to 0.01 mm precision. The gauges were mounted on two reference I-beams. The reference beam supports were at a clear distance > 2.5 m from the test pile head. All test piles were loaded in one cycle. Each increment (25% of the design load) was maintained for a maximum period of two hours or when settlement rate was observed to be less than 0.25 mm per hour. The loading test reaches a maximum load of one and half times the design load. Table 2 shows load increment in the test. Figure 11 shows the load settlement relationships for the four pile load tests.

EXCAVATION OF SITE AND DEWATERING SYSTEM

The site was excavated to the foundation level. The dewatering system was done in the site by surface dewa-

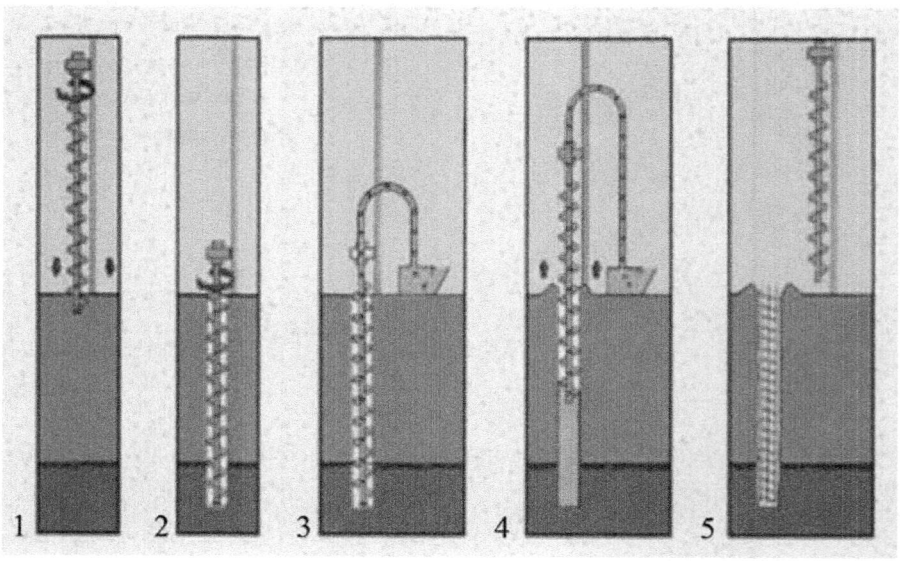

Figure 4: The steps of continuous flight auger (CFA) piles.

Figure 5: The machine used in execution of (CFA) piles.

tering. Figures 12-14 show the steps of excavation process of site and the dewatering system.

ULTIMATE CAPACITY OF PILES USING FIELD LOAD TEST RESULTS

The ultimate capacities of the tested piles were deter-

Figure 6: Fabrication of steel cage at site.

Figure 7: Drilling and concreting of pile by (CFA).

Figure 8: Reinforced steel cage inserted in concrete column by vibration.

Figure 9: The pile cap of tested pile.

Figure 10: Knetledge method for pile test.

Table 2: loading-unloading procedure for pile test

Duration (minute)	60	60	60	180	180	720	15	15	15	15	15	240
% working load	25	50	75	100	125	150	125	100	75	50	25	0

mined from the load test results using different approaches as follows.

Egyptian Code

Egyptian Code recommends that successful pile load test must confirm with the following Equation [7]:

$$S_2/S_1 \geq 1.5 \tag{4}$$

where:

S = settlement of pile;

S_1 = settlement of pile at Q_{all};

S_2 = settlement of pile at 1.25 Q_{all};

Q_{all} = allowable load.

Table 3 shows the calculated allowable load capacity of pile load test using Equation (4) as $S_2/S_1 = 1.5$.

Tangent-Tangent Method

Applying tangenttangent method as used Egyptian Code, a plot is made between stress and the settlement on semi logarithmic scale as shown in **Figure 15** for pile load test #1 [7].

Hansen Method (1963)

Applying Hansen Method the square root of each settlement value from field load test data divided by the corresponding load value is plotted against the settlement as shown in Figure 16 for pile load test #2. Estimation of the ultimate load by Hansen Method is given by the formula [8]:

$$Q_u = \left(2C_1 C_2\right)^{1/2} \tag{5}$$

where:

Q_u = ultimate load capacity;

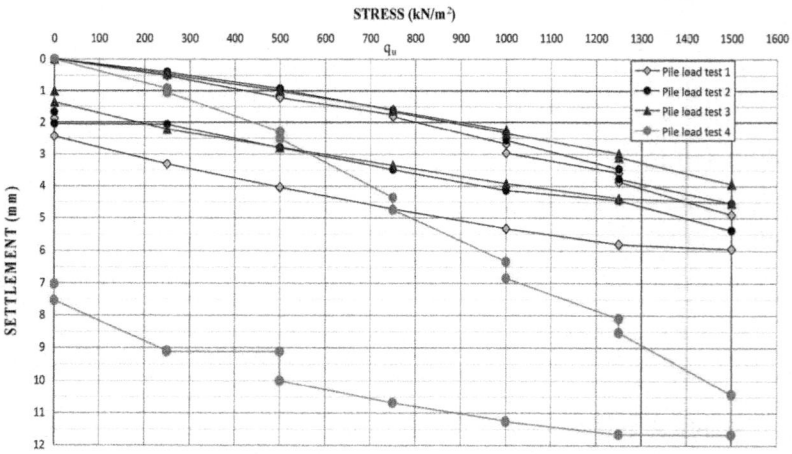

Figure 11: Load-settlement curve for piles load test.

Table 3: Calculated allowable load capacity of pile for pile test

Test No.	Pile test #1		Pile test #2		Pile test #3		Pile test #4	
	settlement of pile (mm)	Q_{all} (kN)	Settlement of pile (mm)	Q_{all} (kN)	settlement of pile (mm)	Q_{all} (kN)	Settlement of pile (mm)	Q_{all} (kN)
S at 1.25 Q_{all}	3.87	1107.8	3.78	1020.5	3.10	1137.3	8.53	1203.3
S at Q_{all}	2.86		2.57		2.35		6.85	

Figure 12: Site after excavation to requirement dept of raft foundation.

C_1 = slope of the best fitting straight line.

C_2 = y-intercept of the straight line.

Chin's Method (1970)

Applying Chin's method (Egyptian Code), a plot is made between settlement divided by corresponding load and the

Figure 13: Remove of pile concrete head.

settlement as shown in **Figure 17** for test pile #3. The inverse slope of the straight line gives the ultimate load as proposed by Chin [9].

Decourt's Extrapolation (1999)

Applying Decourt's Extrapolation by dividing each load by its corresponding settlement and plotting the resulting

Figure 14: Surface dewatering of ground water for the site.

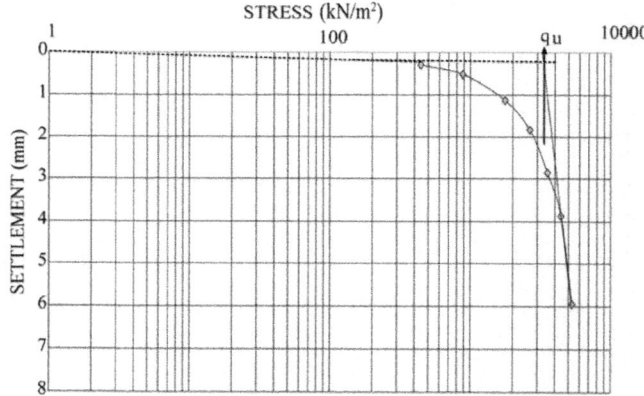

Figure 15: Ultimate pile capacities by tangent-tangent method for working pile load test #1.

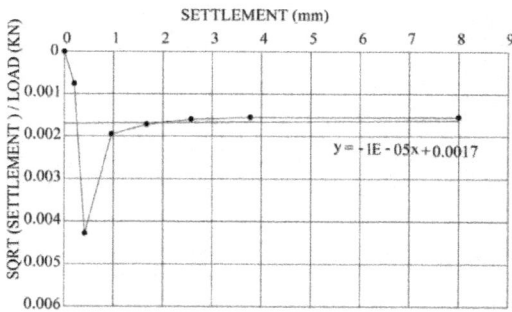

Figure 16: Ultimate pile capacity by Hansen Method for test pile #2.

values against the applied load. A linear regression over the apparent line (last three points) determines a line. Decourt identified the ultimate load as the intersection of this line with load axis as shown in **Figure 18** for working test pile #4 [10].

PROPOSED METHOD FOR DETERMINATION OF ULTIMATE PILE CAPACITY FROM LOAD TEST

The measured settlement of pile is the sum of settlement due to friction and end bearing loads as well as the elastic deformation of the pile itself. The elastic deformation of

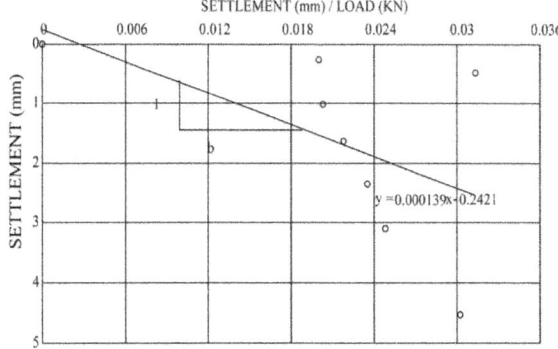

Figure 17: Ultimate pile capacity by Chin's method for test pile #3.

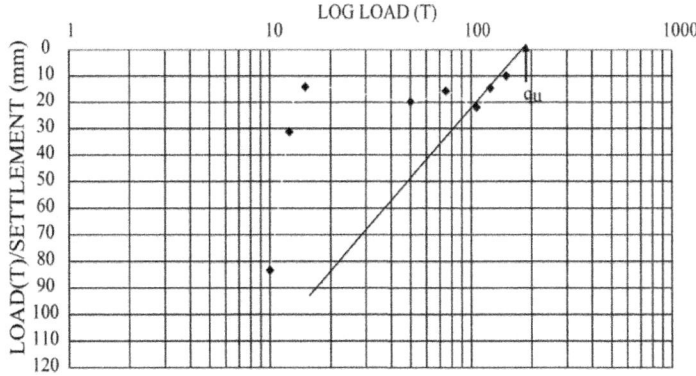

Figure 18: Ultimate pile capacity using Decourt's method for pile load test #4.

pile affects to a great extent the total settlement. However, the elastic deformation depends on the pile length itself and area of pile as well as the quality of concrete. Many authors and Codes such as NAVFAC and Egyptian Code recommended that the settlement at the top of a pile can be divided into three components [11,12].

$$S = S_e + S_b + S_S \tag{6}$$

$$S_e = (Q_b + \alpha Q_s) L / A E_p \tag{7}$$

$$S_b = C_b Q_b / d q_u \tag{8}$$

$$S_s = C_s Q_s / L q_u \tag{9}$$

where:

S_e = elastic deformation of pile;

S_b = settlement of pile base due to the end bearing load;

S_s = settlement of pile base due to the friction load;

Q_b = end bearing load;

Q_s = friction load;

q_u = end bearing resistance;

L = pile depth;

A = area of cross section of pile;

d = pile diameter;

E_p = modulus of elasticity of pile;

$\alpha = 0.67$;

C_b = empirical factor depending on soil type and method of pile installation.

$$C_s = \left[0.93 + 0.16 (L/d)^{1/2} \right] C_b.$$

The proposed method depends on the settlement of pile without taken into consideration the elastic deformation. An empirical formula is presented from the relationship between stress and settlement of pile due to friction and end bearing only after deducting the elastic deformation of pile. However, the determination of ultimate load consists of the steps below:

1) Deducting elastic deformation of pile under each load increment from the measured settlement;

2) Plotting stress settlement curve from field load test data after deducting elastic deformation of pile (Figures 19-22);

3) The following proposed empirical formula has been obtained from the stress settlement mentioned relationships:

$$Q_u = \left[\frac{1}{\alpha * m * y} \right]$$

(10)

where:

Q_u = ultimate load capacity (kN);

m = slope of the trend straight line;

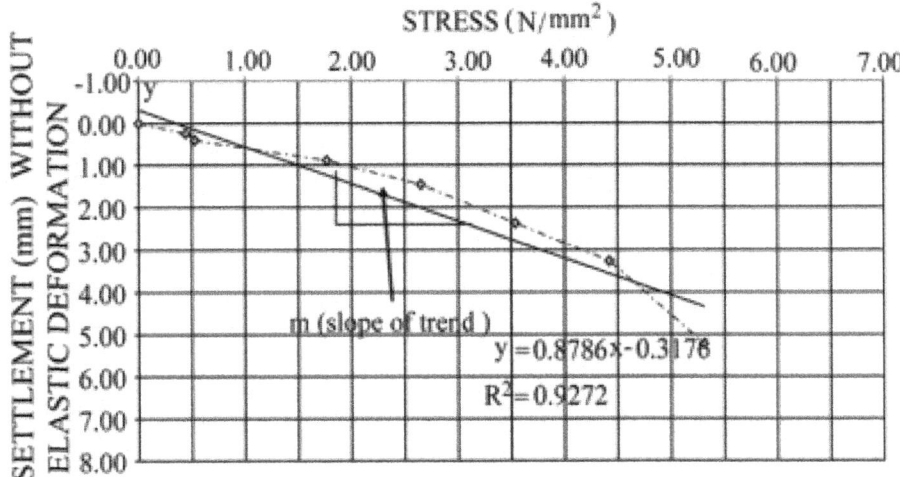

Figure 19: Ultimate pile capacity by using proposed method by the author for test pile #1 at Ep = 19677.40.

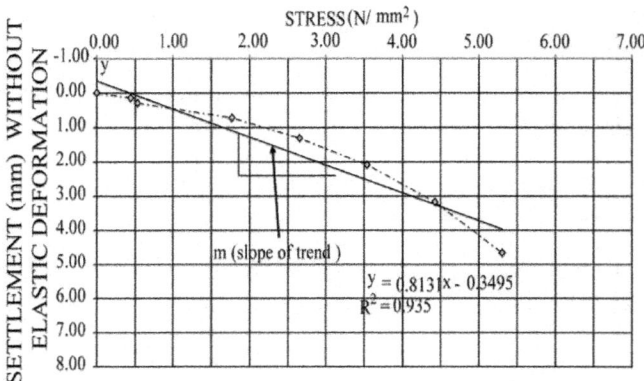

Figure 20: Ultimate pile capacity by using proposed method by the author for test pile #2 at Ep = 19677.40.

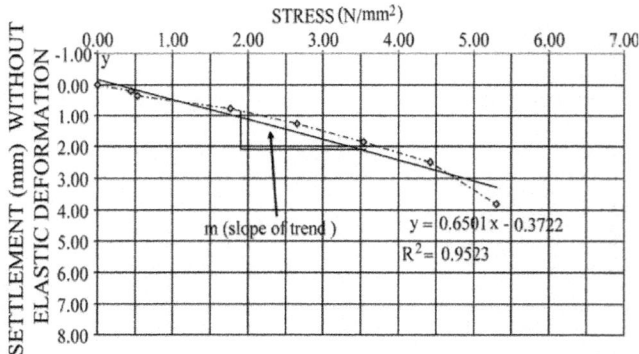

Figure 21: Ultimate pile capacity by using proposed method by the author for test pile #3 at Ep = 19677.40.

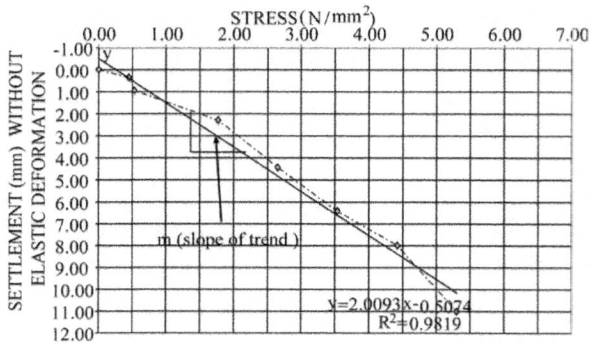

Figure 22: Ultimate pile capacity by using proposed method by the author for test pile #4 at Ep = 19677.40.

y = y-intercept of the straight line (as a value without sign);

α = factor depends on modulus of elasticity of concrete.

The values of the injected concrete elastic modulus (E_p) affect to a great extent the elastic deformation of piles. However, Table 4 shows the values of (E_p) and the corresponding values of the coefficient (α) that to be used in Equation (10).

The obtained results for the ultimate capacity of end bearing piles are considered to be more accurate than other methods. The proposed method appears to give bitter results that agree well with the theoretical predictions. However, Equation (3) is the most suitable and reliable method for predicting the theoretical capacity of piles before execution.

COMPARISON BETWEEN DIFFERENT METHODS OF ULTIMATE PILE CAPACITY DETERMINATION

The calculation of the ultimate capacity of piles and the corresponding factors of safety using the above mention methods are summarized in Table 5. The ultimate capacity obtained by various methods from the pile load test results are shown in Figure 23. However, pile load test #4

Table 4: loading-unloading procedure for pile test

α	0.00159	0.00154	0.00152	0.00150	0.00149
E_p (N/mm^2)	19,677.40	22,000.00	24,099.79	26,030.75	27,828.04

gives lower values due to excess settlement.

The new proposed method by the author for determining the ultimate capacity of end bearing piles appears to give a bitter result that agree well with the analytical predictions. However, Equation (3) is the most suitable and reliable method for predicting the theoretical capacity of piles before execution.

LOAD CARRIED BY END BEARING AND FRICTION ALONG SHAFT

The values of the ultimate pile capacity were taken from **Table 5** and **Figure 3** to evaluate the percentage of friction and end bearing capacity. Based on the above findings, it was found that the percentage of load carried by

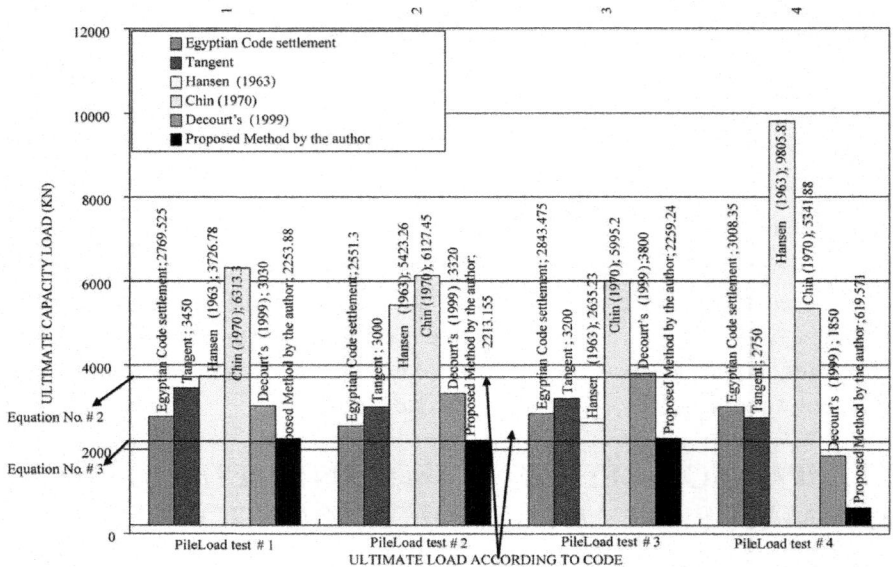

Figure 23: Comparison of ultimate pile loads using different methods.

Table 5: Ultimate Capacity and factor of safety (F.S.) of Pile using different methods

Test No.	Pile test #1		Pile test #2		Pile test #3		Pile test #4	
Metod	Qult (kN)	F.S.	Qult (kN)	F.S.	Qult (kN)	F.S.	Qult (kN)	F.S.
Egyptian Code settlement	2769.53	2.77	2551.3	2.55	2843.48	2.84	3008.35	3.01
Tangent	3450	3.45	3000	3.00	3200	3.20	2750	2.75
Hansen (1963)	3726.78	3.73	5423.26	5.42	2635.23	2.64	9805.81	9.81
Chin (1970)	6313.3	6.31	6127.45	6.13	5995.2	6.00	5341.88	5.34
Decourt's (1999)	3030	3.03	3320	3.32	3800	3.80	1850	1.85
Proposed Method by the author at Ep = 19677.40	2253.88	225	2213.155	2.22	2259.24	2.26	619.571	0.62

Table 6: Percentage of ultimate load carried by end bearing and friction

Test No.	Pile test #1		Pile test #2		Pile test #3		Pile test #4	
Metod	Skin friction %	End bearing %	Skin friction %	End bearing %	Skin friction %	End bearing %	Skin friction %	End bearing %
Egyptian Code settlement	39.75	60.25	45.71	54.29	42.85	57.15	49.87	50.13
Tangent	39.75	60.25	45.71	54.29	42.85	57.15	49.87	50.13
Hansen (1963)	36.80	63.20	25.29	74.71	52.04	47.96	13.98	86.02
Chin (1970)	21.72	78.28	22.38	77.62	22.87	77.13	25.67	74.33
Decourt's (1999)	45.26	54.74	41.30	58.70	36.09	63.91	74.12	25.88
Proposed Method by the author at Ep = 19677.40	45.58	54.42	46.42	53.58	45.47	54.53	68.17	31.83

friction along the pile shaft and the end bearing are shown in the **Table 6**. However, it can be shown that the proposed method gives the percentage of friction load carried by the shaft along the pile length to be about 46% of total load while the pile load carried by the end bearing is 54% of total load. In addition, the proposed method gives bitter results that agree with Equation (3).

CONCLUSIONS

From the test results the following conclusions are arrived:

1) A new proposed method to calculate the ultimate capacity of pile from pile load test is presented. The method is based on the relationship between the stress and settlement after deducting the elastic deformation of pile.

2) The proposed method for determining the ultimate capacity appears to give bitter results that agrees well with the theoretical predictions. The proposed method gives bitter results that agree with theoretical predictions.

3) The percentage of friction load carried by the shaft along the pile length is about 46% of total load, while the pile load carried by the end bearing is 54% of total load.

The proposed method is easier, quicker and more reliable.

ACKNOWLEDGEMENTS

The author would like to acknowledge the Fetih Construction Company for their valuable assistance.

REFERENCES

1. T. Y. C. Ir, C. M. G. Chow and S. B. Partners, "Design & Construction of Bored Pile Foundation," Geotechnical Course for Pile Foundation Design & Construction, Ipoh, 29-30 September 2003, pp. 1-74.

2. B. Dan, D. D. Steven, W. T. Robert and A. L. Carlos "Design and Construction of Continuous Flight Auger (CFA) Piles," Geotechnical Engineering Circular No. 8. April Office of Technology Application Office of Engineering/Bridge, Division Federal Highway Administration, US Department of Transportation, Washington DC, 2007, pp. 33-65.

3. Akbar, S. Khilji, S. B. Khan, M. S. Qureshi and M. Sattar, "Shaft Friction of Bored Piles in Hard Clay," Pakistan Journal of Engineering and Applied Science, Vol. 3, No. 3, 2008, pp. 54-60.

4. K. Ishihara, "Recent Advances in Pile Testing and Diaphragm Wall

Construction in Japan," Geotechnical Engineering Journal of the SEAGS & AGSSEA, Vol. 41, No. 3, 2010, pp. 1-43.

5.   M. Suman and Y. Noriyuki, "End Bearing Capacity of Tapered Piles in Sands using Cavity Expansion Theory," Memoirs of the Faculty of Engineering, Kyushu University, Vol. 71, No. 4, 2011, pp. 77-99.

6.   W. N. Elsamee, "Evaluation of the Ultimate Capacity of Friction Piles," Engineering, Vol. 4 No. 11, 2012, pp. 778-789. doi:10.4236/eng.2012.411100

7.   Egyptian Code, "Soil Mechanics and Foundation," Organization, Cairo, 2005.

8.   J. B. Hansen, "Discussion on Hyperbolic Stress-Strain Response, Cohesive Soils," Journal for Soil Mechanics and Foundation Engineering, Vol. 89, 1963, pp. 241-242.

9.   F. K. Chin, "Estimation of the Ultimate Load of Piles from Tests Not Carried to Failure," Proceedings of Second Southeast Asian Conference on Soil Engineering, Singapore, 11-15 June 1970, pp. 81-92.

10. D. M. Dewaikar and M. J. Pallavi, "Analysis of Pile Load Tests Data" Journal of Southeast Asian Geotechnical Society, Vol. 89, No. 4, 2000, pp. 27-39.

11. M. Das Braja, "Principle of Foundation Engineering," 7th Edition, USA Cengage Brain.com, 2011.

12. NAVFAC, DM-7.2, "Foundations and Earth Structures," Design Manual, Department of the Navy Facilities Engineering Command, Alexandria, 1982.

Chapter 5

FLEXURAL BEHAVIOR OF LATERALLY LOADED TAPERED PILES IN COHESIVE SOILS

Musab Aied Qissab

Department of Civil Engineering, Al-Nahrain University, Baghdad, Iraq

ABSTRACT

In this paper, the flexural behavior of laterally loaded tapered piles in cohesive soils is investigated. The exact solution for the governing differential equation of the problem is obtained based on the beam-on-elastic foundation approach in which the soil reaction on the pile is related directly to the pile lateral deflection. In this investigation, the modulus of subgrade reactions is assumed to be constant along the pile depth. Parametric study through numerical examples is carried out to prove the validity and accuracy of the obtained results. In general, the derived displacement field can be used to study pile response in multilayered soil profiles by subdividing the pile into a number of elements. It is found that tapered piles show stiffer behavior than that for prismatic ones having the same material volume with an optimum stress distribution along the pile depth. Accordingly, tapered piles are more efficient and economic than those having the same material volume. Verification is also carried out for the obtained results through finite element analysis and the selected number of elements gives a very good agreement for lateral deflection and a larger number of elements is required to obtain better results for bending moment because of moment loss resulting from the lack of shear diagram.

INTRODUCTION

Piles are widely used to support structures not limited to bridges, high rise buildings and offshore structures which are subjected toaxial and lateral loads resulting from different sources. Tapered piles, as special cases, have received great attention at present due to their good performance in resisting loads in comparison to that for prismatic ones because of the optimum material distribution with respect to loading intensity. Most of the available analysis and design guidelines lay more emphasis on prismatic piles over tapered piles despite of the economical advantage of the latter. Tapered piles are not widely used as a design option because of the limit knowledge about their behavior under different loading types in comparison to the prismatic piles.

There are a number of studies concerning the behavior of individual piles. Wei [1] studied experimentally the static behavior of piles in cohesionless soils under the effect of axial, lateral, and cyclic loads. Two sets of tests with three types of geometries including a prismatic pile in dry sandy soil were conducted to study their beha- vior. The results of the study confirmed the efficiency of tapered piles over the prismatic ones having the same material input.

Horvath et al. [2] investigated experimentally the behavior of tapered tube piles under axial, uplift, and lateral loads in sand. The experimental program was mainly conducted for one of the larger transportation projects for the major renovation and expansion of John F. Kennedy International Airport in New York City to verify the performance of these piles. It was established from the experimental results that taper-tube piles are successfully resist the entire spectrum of axial and lateral loads that is normally encountered in transportation engineering.

Shankar et al. [3] developed a procedure to predict the flexural behavior of axially loaded and laterally loaded tapered piles embedded in liquefaction-induced laterally spread soils. The problem was analyzed by using the modulus of subgrade reaction approach based on Winkler type soil model. The resulting governing equation to solve the flexural behavior of the pile with the specified boundary conditions was solved by using finite difference technique. The use of tapered piles was found beneficial in liquefaction-induced laterally spreading soils as the maximum bending moment developed due to drag force is less especially when the applied axial force is much lower than the critical load.

Zhan et al. [4] studied the load capacity behavior of two series of axially loaded tapered piles in sand by using finite element method. It was observed from the numerical analysis that the shaft resistance increasing with the tapered angle with an increase of (12%) over that of the straight-side piles at an optimum tapered angle. It was concluded that tapered piles are more suitable for floating pile foundations.

STATEMENT OF THE PROBLEM

The tapered pyramidal pile shown in Figure 1 of length (L) and embedded in a homogeneous cohesive soil layer. is subjected at its head to a lateral concentrated load (Q) and a bending moment (M). The governing differential equation of the above problem was given by Hetenyi [5] for beams on elastic foundation with variable flexural rigidity as follows:

$$\frac{d^2}{dz^2}\left[EI_{(z)}\frac{d^2 y}{dz^2}\right]+k_z y = 0$$

(1)

Or

$$E\left[I_{(z)}\frac{d^4 y}{dz^4}+2\left(\frac{dI_{(z)}}{dz}\right)\frac{d^3 y}{dz^3}+\left(\frac{d^2 I_{(z)}}{dz^2}\right)\frac{d^2 y}{dz^2}\right]+k_z y = 0$$

(2)

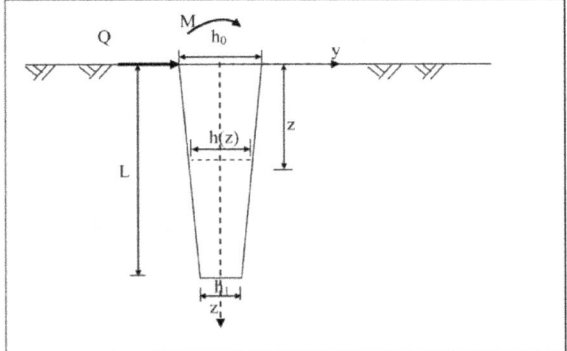

Figure 1: The considered pile configuration.

where k_z is the modulus of subgrade reaction $= Kb(z)$, and E is the modulus of elasticity of the pile material. The moment of inertia I(z) can be given by:

$$I_{(z)} = \frac{b_{(z)}\left(h_{(z)}\right)^3}{12} = \frac{b_0 h_0^3}{12}(1-cz)^4 = I_0(1-cz)^4$$

(3)

in which:

$b_{(z)} = b_0(1-cz)$, $h_{(z)} = h_0(1-cz)$; b_0, h_0 are the section dimensions of the pile at z = 0; $c = (h_0 - h_1)/(h_0 L)$;

h_1 = the pile dimension in the direction of y at z = L; and I_0 is the moment of inertia at z = 0.

Substituting Equation (3) into (2) and assuming that $\bar{z} = (1-cz)$ and after simplifications, the following equation can be obtained:

$$c^4 \bar{z}^4 \frac{d^4 y}{d\bar{z}^4} + 8c^4 \bar{z}^3 \frac{d^3 y}{d\bar{z}^3} + 12c^4 \bar{z}^2 \frac{d^2 y}{d\bar{z}^2} + \frac{k_z}{EI_0} y = 0$$

(4)

The solution of Equation (4) can be expressed as follows:

$$y(z) = A_1 \bar{z}^{(a_1 + b_1(i))} + A_2 \bar{z}^{(a_2 + b_2(i))} + A_3 \bar{z}^{(a_3 + b_3(i))} + A_4 \bar{z}^{(a_4 + b_4(i))}$$

(5)

or

$$y(z) = A_1 \bar{z}^{(a_1)} e^{i(b_1 \ln(\bar{z}))} + A_2 \bar{z}^{(a_2)} e^{i(b_2 \ln(\bar{z}))} + A_3 \bar{z}^{(a_3)} e^{i(b_3 \ln(\bar{z}))} + A_4 \bar{z}^{(a_4)} e^{i(b_4 \ln(\bar{z}))}$$

(6)

where

$$\left.\begin{aligned}
a_1 &= \frac{1}{2}(-1-a), b_1 = -\frac{\bar{\beta}}{a} \\
a_2 &= \frac{1}{2}(-1+a), b_2 = \frac{\bar{\beta}}{a} \\
a_3 &= \frac{1}{2}(-1-a), b_3 = \frac{\bar{\beta}}{a} \\
a_4 &= \frac{1}{2}(-1+a), b_4 = -\frac{\bar{\beta}}{a}
\end{aligned}\right\}$$

(7)

$$a = \sqrt{\frac{5}{2}\left[1 + \sqrt{1 + \left(\frac{4}{5}\bar{\beta}\right)^2}\right]}, \quad \bar{\beta} = \sqrt{\frac{\beta^4}{c^4} - 1}, \quad \beta = \frac{k_z}{EI_0}, \quad \text{and} \quad \bar{z} = (1 - cz)$$

(8)

The constants A_1, A_2, A_3, and A_4 can be determined by applying the boundary conditions as follows:

at $(z = 0)$

$$\frac{d^2 y}{dz^2} = \frac{M}{EI_0}$$

(9)

$$E\left[I_{(z)} \frac{d^3 y}{dz^3} + \frac{dI_{(z)}}{dz} \frac{d^2 y}{dz^2} \right] = Q$$

(10)

or

$$\left[(\bar{z}) \frac{d^3 y}{dz^3} - 4c \frac{d^2 y}{dz^2} \right] = \frac{Q}{EI_0}$$

(11)

at $(z = L)$

$$\frac{d^2 y}{dz^2} = 0$$

(12)

$$\left[(\bar{z}) \frac{d^3 y}{dz^3} - 4c \frac{d^2 y}{dz^2} \right] = 0$$

(13)

Applying the above boundary conditions, the following equations can be obtained:

at $(z = 0)$

$$\left(\bar{z}^{a_1} D_{21} - 2a_1 c D_{11} + a_1 (a_1 - 1) c^2 D_{01} \right) A_1 + \left(\bar{z}^{a_1} D_{22} - 2a_1 c D_{12} + a_1 (a_1 - 1) c^2 D_{02} \right) A_2$$
$$+ \left(\bar{z}^{a_2} D_{21} - 2a_2 c D_{11} + a_2 (a_2 - 1) c^2 \right) A_3 + \left(\bar{z}^{a_2} D_{21} - 2a_2 c D_{11} + a_2 (a_2 - 1) c^2 \right) A_4 = M / (EI_0)$$

(14)

$$\left(\bar{z} D_{311} - 4c D_{211} \right) A_1 + \left(\bar{z} D_{312} - 4c D_{212} \right) A_2 + \left(\bar{z} D_{321} - 4c D_{221} \right) A_3 + \left(\bar{z} D_{322} - 4c D_{222} \right) A_4 = \frac{Q}{EI_0}$$

(15)

at $(z = L)$

$$\left(\bar{z}^{a_1} D_{21} - 2a_1 c \bar{z}^{(a_1 - 1)} D_{11} + a_1 (a_1 - 1) c^2 \bar{z}^{(a_1 - 2)} D_{01} \right) A_1$$
$$+ \left(\bar{z}^{a_1} D_{22} - 2a_1 c \bar{z}^{(a_1 - 1)} D_{12} + a_1 (a_1 - 1) \bar{z}^{(a_1 - 2)} c^2 D_{02} \right) A_2$$
$$+ \left(\bar{z}^{a_2} D_{21} - 2a_2 c \bar{z}^{(a_2 - 1)} D_{11} + a_2 (a_2 - 1) c^2 \bar{z}^{(a_2 - 2)} D_{01} \right) A_3$$
$$+ \left(\bar{z}^{a_2} D_{22} - 2a_2 c \bar{z}^{(a_2 - 1)} D_{12} + a_2 (a_2 - 1) c^2 \bar{z}^{(a_2 - 2)} D_{02} \right) A_4 = 0$$

(16)

$$\left(\bar{z} D_{311} - 4c D_{211} \right) A_1 + \left(\bar{z} D_{312} - 4c D_{212} \right) A_2 + \left(\bar{z} D_{321} - 4c D_{221} \right) A_3 + \left(\bar{z} D_{322} - 4c D_{222} \right) A_4 = 0$$

(17)

where:

$$\left. \begin{array}{l} D_{01} = \sin\left(\frac{\bar{\beta}}{a} \ln(\bar{z}) \right) \\[2mm] D_{02} = \cos\left(\frac{\bar{\beta}}{a} \ln(\bar{z}) \right) \\[2mm] D_{11} = -\frac{\bar{\beta}}{a} \left(\frac{c}{z} \right) \cos\left(\frac{\bar{\beta}}{a} \ln(\bar{z}) \right) \\[2mm] D_{12} = \frac{\bar{\beta}}{a} \left(\frac{c}{z} \right) \sin\left(\frac{\bar{\beta}}{a} \ln(\bar{z}) \right) \\[2mm] D_{21} = -\gamma^2 \sin\left(\frac{\bar{\beta}}{a} \ln(\bar{z}) \right) - \frac{\bar{\beta}}{a} \left(\frac{c}{z} \right)^2 \cos\left(\frac{\bar{\beta}}{a} \ln(\bar{z}) \right) \\[2mm] D_{22} = -\gamma^2 \cos\left(\frac{\bar{\beta}}{a} \ln(\bar{z}) \right) + \frac{\bar{\beta}}{a} \left(\frac{c}{z} \right)^2 \sin\left(\frac{\bar{\beta}}{a} \ln(\bar{z}) \right) \end{array} \right\}$$

(18)

and

$$D_{211} = \overline{z}^{a_1} D_{21} - 2a_1 c\overline{z}^{(a_1-1)} D_{11} + a_1(a_1-1)c^2\overline{z}^{(a_1-2)} D_{01}$$

$$D_{212} = \overline{z}^{a_1} D_{22} - 2a_1 c\overline{z}^{(a_1-1)} D_{12} + a_1(a_1-1)c^2\overline{z}^{(a_1-2)} D_{02}$$

$$D_{221} = \overline{z}^{a_2} D_{21} - 2a_2 c\overline{z}^{(a_2-1)} D_{11} + a_2(a_2-1)c^2\overline{z}^{(a_2-2)} D_{01}$$

$$D_{222} = \overline{z}^{a_2} D_{22} - 2a_2 c\overline{z}^{(a_2-1)} D_{12} + a_2(a_2-1)c^2\overline{z}^{(a_2-2)} D_{02}$$

$$D_{311} = \overline{z}^{a_1} D_{31} - 3a_1 c\overline{z}^{(a_1-1)} D_{21} + 3a_1(a_1-1)c^2\overline{z}^{(a_1-2)} D_{11} - a_1(a_1-1)(a_1-2)c^3\overline{z}^{(a_1-3)} D_{01}$$

$$D_{312} = \overline{z}^{a_1} D_{32} - 3a_1 c\overline{z}^{(a_1-1)} D_{22} + 3a_1(a_1-1)c^2\overline{z}^{(a_1-2)} D_{12} - a_1(a_1-1)(a_1-2)c^3\overline{z}^{(a_1-3)} D_{02}$$

$$D_{321} = \overline{z}^{a_2} D_{31} - 3a_2 c\overline{z}^{(a_2-1)} D_{21} + 3a_2(a_2-1)c^2\overline{z}^{(a_2-2)} D_{11} - a_2(a_2-1)(a_2-2)c^3\overline{z}^{(a_2-3)} D_{01}$$

$$D_{322} = \overline{z}^{a_2} D_{32} - 3a_2 c\overline{z}^{(a_2-1)} D_{22} + 3a_2(a_2-1)c^2\overline{z}^{(a_2-2)} D_{12} - a_2(a_2-1)(a_2-2)c^3\overline{z}^{(a_2-3)} D_{02}$$

(19)

in which,

$$D_{31} = \gamma^3 \cos\left(\frac{\overline{\beta}}{a}\ln(\overline{z})\right) - 2\left(\frac{\overline{\beta}}{a}\right)^2\left(\frac{c}{\overline{z}}\right)^3 \sin\left(\frac{\overline{\beta}}{a}\ln(\overline{z})\right) - \gamma^2\left(\frac{c}{\overline{z}}\right)\sin\left(\frac{\overline{\beta}}{a}\ln(\overline{z})\right) + 2\frac{\overline{\beta}}{a}\left(\frac{c}{\overline{z}}\right)^3 \cos\left(\frac{\overline{\beta}}{a}\ln(\overline{z})\right)$$

$$D_{32} = -\gamma^3 \sin\left(\frac{\overline{\beta}}{a}\ln(\overline{z})\right) - 2\left(\frac{\overline{\beta}}{a}\right)^2\left(\frac{c}{\overline{z}}\right)^3 \cos\left(\frac{\overline{\beta}}{a}\ln(\overline{z})\right) - \gamma^2\left(\frac{c}{\overline{z}}\right)\cos\left(\frac{\overline{\beta}}{a}\ln(\overline{z})\right) + 2\frac{\overline{\beta}}{a}\left(\frac{c}{\overline{z}}\right)^3 \sin\left(\frac{\overline{\beta}}{a}\ln(\overline{z})\right)$$

(20)

$$\gamma = \left(\frac{\overline{\beta}}{a}\right)\left(\frac{c}{\overline{z}}\right)$$

(21)

NUMERICAL EXAMPLES

In this paper, the behavior of two groups of piles (with square cross-section) is investigated. The geometry, loading condition and materials constants for each group are presented in Figure 2 andFigure 3.

The lateral deflection, the distribution of shear force and bending moment along the pile shaft are given in Figures 4-6.

The results of group No. 2 are given in Figures 7-9. Pile (S2) in this group has the same material volume for pile (T3).

It can be observed from Figure 4 and Figure 7 that tapered piles show stiffer behavior than prismatic ones having the same material volume. The deflection decrease for a tapering angle (0.955°) (pile T1) is found to be

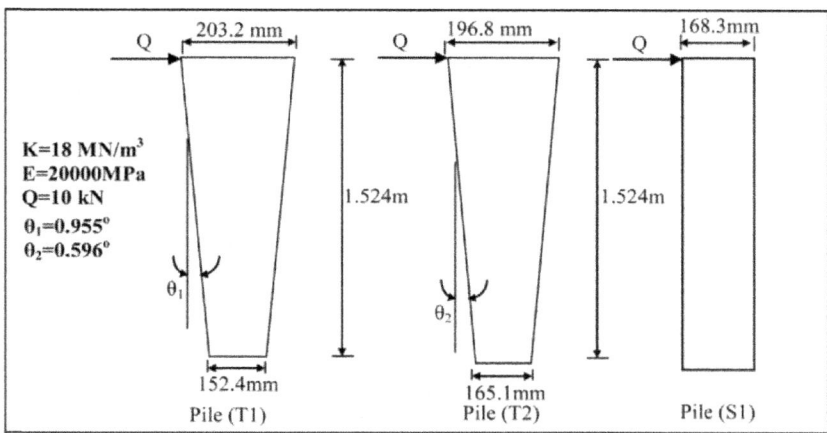

Figure 2: Pile group No. 1.

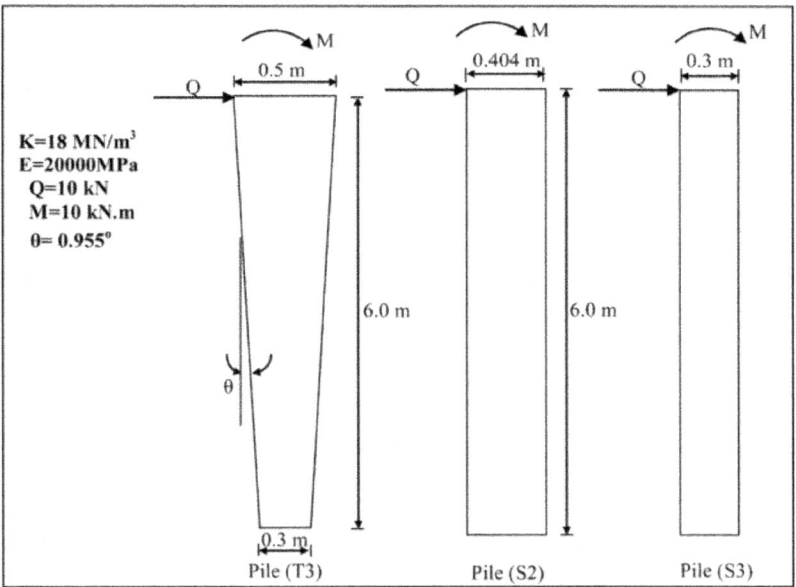

Figure 3: Pile group No. 2.

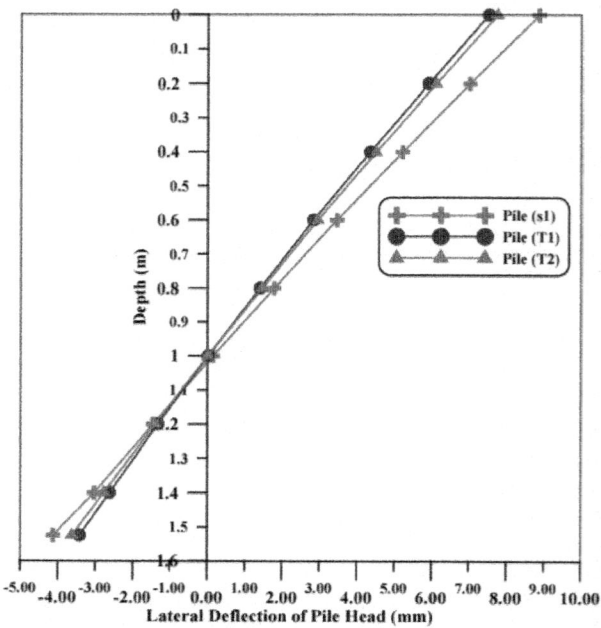

Figure 4: Lateral deflection of pile group No. 1.

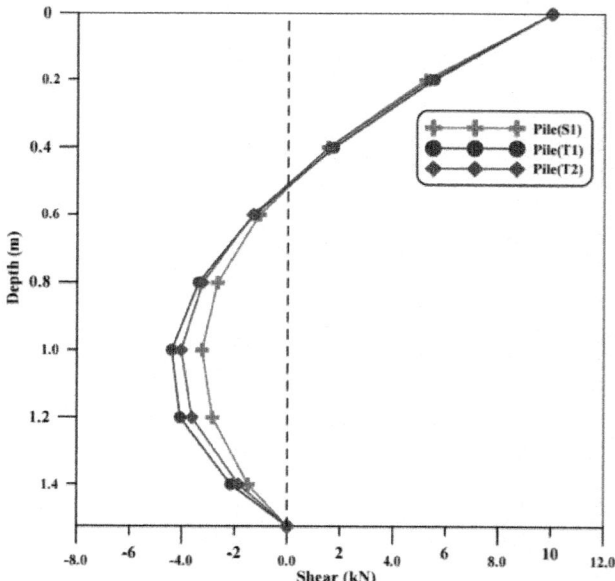

Figure 5: Shear force distribution for pile group No. 1.

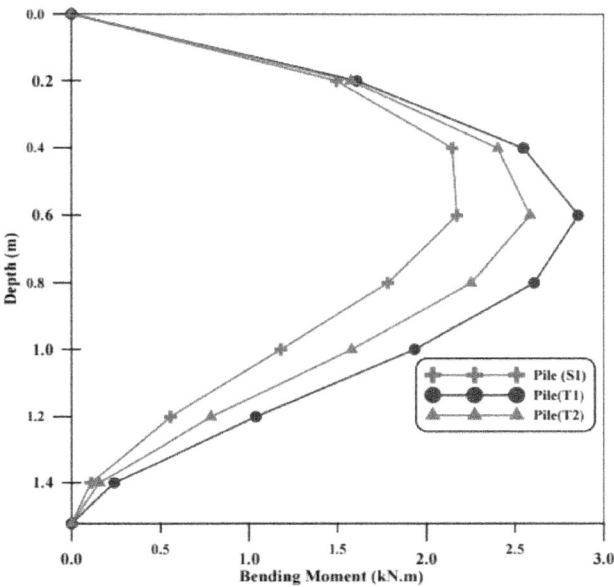

Figure 6: Moment distribution for pile group No. 1.

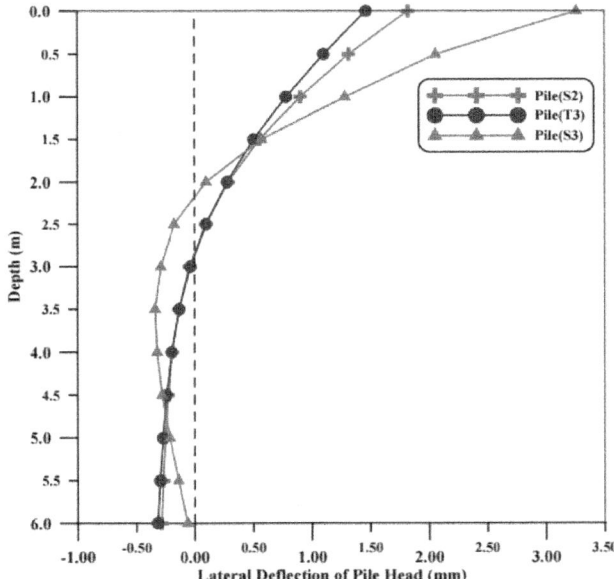

Figure 7: Lateral deflection for pile group No. 2.

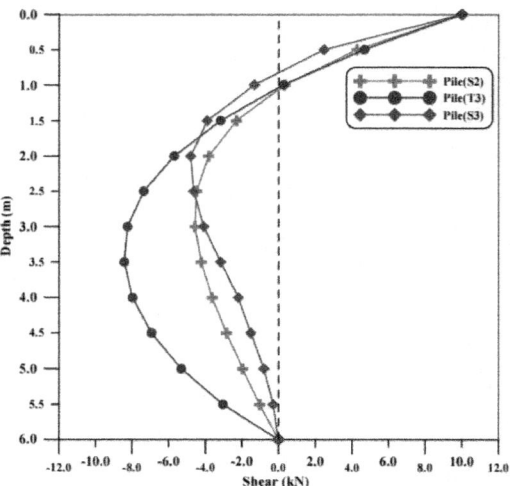

Figure 8: Shear force distribution for pile group No. 3.

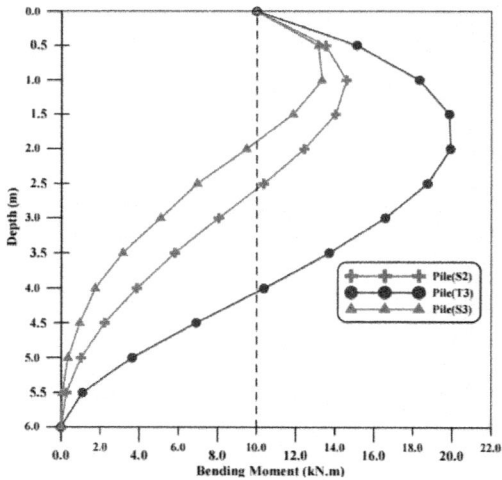

Figure 9: Moment distribution for pile group No. 3.

(17.84%) for group No. 1 and (22%) for group No. 2 (pile T3) for the same material volume. Also, it can be noted that the point of maximum bending moment is located approximately at the upper third of the pile depth at which the cross-sectional area of the tapered piles is larger than that for prismatic ones having the same material volume. This gives a more efficient material distribution.

VERIFICATION OF THE PRESENT SOLUTION

To verify the obtained results from the derived equations and because there is no laboratory studies that concerned with laterally loaded tapered piles in cohesive soils, the finite element method is used to analyze piles T1 and T3 of the two groups. Each pile is subdivided into a number (10) of straight elements as shown in Figure 10 and a computer program is used to solve the problem.

It can be noted from the Figure 11 and Figure 13 that the lateral deflection curves obtained from the two solutions are identical. On the other hand, the moment distribution curves shown in Figure 12 and Figure 14 obtained by the finite element method give a lower-bound solution for the bending moment. This is due to the lack in shear diagram resulting from subdividing the pile into a number of straight elements which cause a loss in bending moment (the area of the shear diagram). To obtain better results for the bending moment, finer divisions for the pile should be used. This verifies the accuracy and efficiency of the proposed solution in comparison to the finite element method.

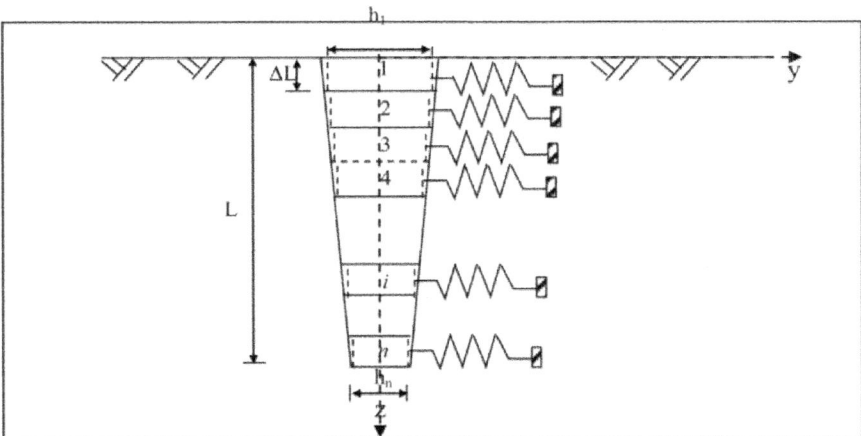

Figure 10: The proposed finite element model of the tapered pile.

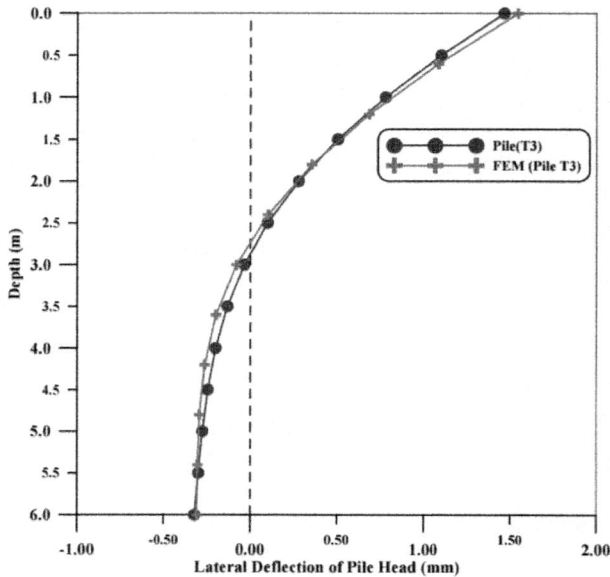

Figure 11: Lateral deflection for piles T4 and S4.

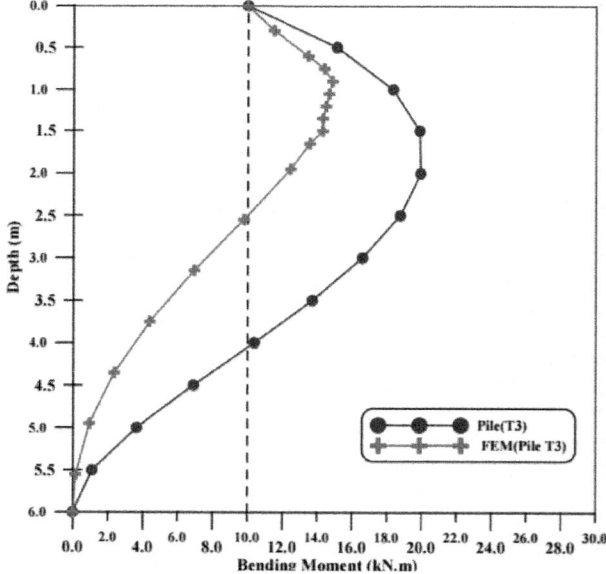

Figure 12: Moment distribution for piles T4 and S4.

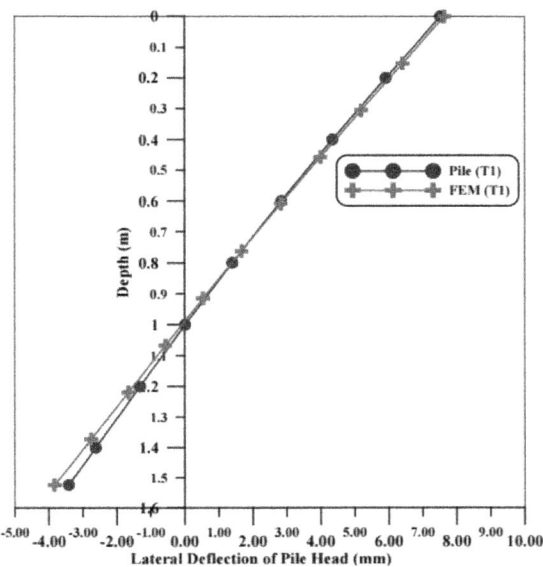

Figure 13: Lateral deflection for pile T1.

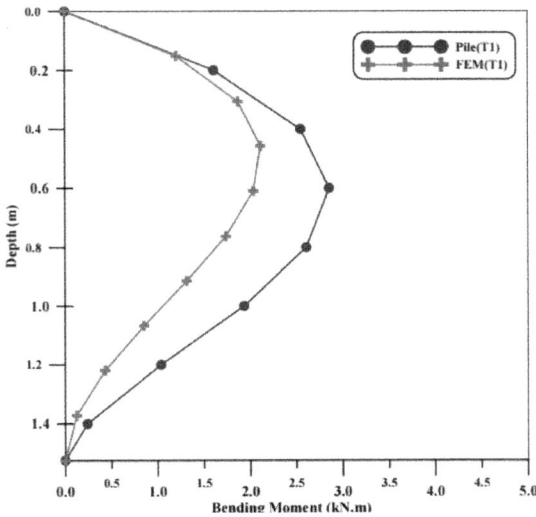

Figure 14: Moment distribution for pile T1.

CONCLUSION

In this paper, an exact solution of the differential equation for a pyramidal tapered pile with square cross-sec- tional area in cohesive soils has been obtained based on beam-on-elastic foundation theory assuming constant value for the subgrade reaction coefficient. It is clear from the presented results that tapered piles show stiffer behavior than that for prismatic ones having the same material volume. It is found that the decrease in deflection for a tapering angle 0.955° is in the range of 17.84% - 22% for the studied cases. The value of the maximum bending moment for tapered piles is found to be larger than that for prismatic piles. The increase in maximum bending moment for tapered piles is proportional to the increase in cross-section size which gives approximately constant bending stress. As a result, tapered piles are more efficient and economic than those having the same material volume.

REFERENCES

1. Wei, J.Q. (1998) Experimental Investigation of Tapered Piles. M.Sc. Thesis, University of Western Ontario, London, Canada.

2. Horvath, J.S., Trochalides, T., Burns, A. and Merjan, S. (2004) A New Type of Tapered Steel Pipe Pile for Transportation Applications. Geo-Trans, American Society of Civil Engineers Geo-Institute, California.

3. Shanker, K., Basudhar, P.K. and Patra, N.R. (2008) Flexural Response of Tapered Piles in Liquefied Soils. The 12th International Conference of International Association for Computer Methods and Advances in Geomechanics (IACMAG), Goa.

4. Zhan, Y., Wang, H. and Liu, F. (2012) Numerical Investigation on Load Capacity Behavior of Tapered Pile Foundations. EJGE, 17, 1969-1980.

5. Hetenyi, M. (1946) Beams on Elastic Foundations. University of Michigan Press, Ann Arbor.

Chapter 6

NUMERICAL STUDY OF PILES GROUP UNDER SEISMIC LOADING IN FRICTINAL SOIL— INCLINATION EFFECT

Fadi Hage Chehade[1], Marwan Sadek[2], Douaa Bachir[3]

[1]Civil Engineering Department, Doctotral School of Sciences and Technology, Lebanese University, University Institute of Technology (Saida) & Modeling Center, Beirut, Lebanon

[2]Laboratory of Civil Engineering and GeoEnvironment, University of Lille I Sciences and Technology, Villeneuve d'Ascq, France

[3]Numerical Center, Doctoral School of Science and Technology, Lebanese University, Beirut, Lebanon

ABSTRACT

Recent devastating earthquakes in some countries, such as Pakistan, Turkey, Algeria and China, call to the mind the high risk exposure of Lebanon which is located over an active seismic zone. Many experts shared the view that major seismic event may occur in Lebanon in the future. Moreover, many earthquakes, of low magnitudes between three and four, have been registered in Lebanon during 2008. These events have increased the anxiety of Lebanese people because of the poor quality of the constructions and their behavior under moderate or severe earthquake events. The efficient way to minimize seismic effects, material and human losses, is the prevention. The system piles-foundation is an appropriate way and widely used to ensure the stability of constructions when subjected to seismic excitation. It seems necessary to study the interaction of pile-foundation-pile-cap-structure in the case of non linear soil behavior and the interface pile-soil. The study will be also conducted by using measures recorded during real earthquakes for example in Turkey (Kocaeli, 1999). In this paper, we present a numerical modeling of the interaction of using FLAC3D software. According to soil behavior and pile inclination, parametric studies are also performed. The analysis of the results could give the better piles group configuration in order to minimize the seismic effect on the structures.

INTRODUCTION

Recent devastating earthquakes in some countries, such as Pakistan, Turkey, Algeria and China, call to the mind the high risk exposure of Lebanon which is located over an active seismic zone. There are some faults in the country represented respectively by major and secondary ones. Many geologic experts shared the view that major seismic event could occur in Lebanon in the future. Moreover, many earthquakes, of low magnitudes between three and four, have been registered in Lebanon during 2008. These events have increased the anxiety of Lebanese people because of the poor quality of the constructions and their behavior that could undergo in the case of moderate or severe earthquake events. The efficient way to minimize seismic effects both material and human losses, is the prevention and in particular it is very important to enhance the foundations of the constructions.

Piles are used as foundation elements for structures located in seismic areas. They provide stability, but may be acquired by efforts that exceed their carrying capacity. These efforts are dangerous for piles installed in soils with low fundamental frequencies, as they amplify the seismic motion of the soil endangering the stability of these structures and their functioning.

Analysis of the seismic response of soil-pile-structure systems constitutes a complex problem in earthquake engineering. In addition to post-earthquake investigationsanalytical and numerical analyses show that the damage of piles in seismic area is mainly attributed to the kinematic interaction between piles and soils or (and) to the inertial interaction between the superstructure and the pile foundation which may cause foundation damages, in particular at the pile-cap connection.

Due to the complexity of the problem of interaction soil-pile-structure and the strong coupling between the elements of foundation and structure, it is necessary to conduct a comprehensive analysis of this problem. Most researches have been conducted within the framework of the elasticity, considering the link between piles and soil as rigid. In the case of strong seismic loading, the nonlinearities of soil can play an important role in modifying the state of the interface soil-pile causing a strong damping of the seismic energy injected into the structure. Using advanced computing resources, the consideration of the nonlinearities of the soil and the structure becomes possible in a comprehensive approach. Non linear full 3D analyses considering the soil, piles and the superstructure are still limited. Such studies were conducted in the linear domain ([1-5]) to analyse the influence of micropiles inclination and boundaries conditions on the seismic behaviour of the soil-micropile structure system. Gerolymos et al. [6] used a full 3D finite element analysis to study the seismic performance of inclined piles assuming a linear behaviour of the soil and the structure.

The present paper is focused on a full 3D coupled modelling of the soil-pile-superstructure interaction under seismic loading considering the elastoplastic behaviour of the soil material. Analysis is performed using the FLAC3D [7] program under real earthquake records (Kocaeli, 1999). Soil plasticity is investigated in the case of frictional soils where the soil behaviour is described using the simple and popular non-associated Mohr-Coulomb criterion largely used in engineering practice. The last part discusses the efficiency of inclined piles in seismic zones. Using inclined piles in seismic zones is generally not recommended by international codes, especially when piles are anchored in hard substrata. However, the analysis of the Loma Prieta earthquake ([8]) and Kobe ([9,10]) showed that structures based on inclined piles were less affected or damaged than other structures.

NUMERICAL ANALYSIS OF SOIL-PILE-STRUCTURE SYS-TEM (ELASTIC SOIL)

Problem Definition

The model that has been analyzed consists of a group of four piles of 1m diameter and 10 m length, embedded in a homogeneous soil layer of 15 m of thickness (Figure 1). The piles are fixed in a cap of 1m thick with no contact with the ground, and supporting a superstructure. The spacing between piles is $S = 3.75D_p$ (D_p: is the diameter of the pile). The behaviour of the soil-pile-structure system is firstly assumed to be elastic with Rayleigh damping. The superstructure is modelled by a single degree of freedom system, consisting of a column of height $H_{st} = 1$ m and a concentrated mass $m_{st} = 100$ tons placed on the top of column. The fundamental frequency of the soil layer is equal to $f_1 = 0.67$ Hz. The rigidity (K_{st}) and the frequency of the superstructure (f_{st}), assumed fixed at its base, are calculated using the following expression:

$$K_{st} = \frac{3\left(E_{st}I_{st}\right)}{\left(H_{st}\right)^3}$$

$$f_{st} = \frac{1}{2\pi}\sqrt{\frac{K_{st}}{m_{st}}}$$

(1)

Then, $K_{st} = 86400$ kN/m and $f_{st} = 1.48$ Hz.

The mesh used is shown in Figure 2. A refinement is located around the piles and the area near the superstructure where the inertial forces induce high stresses. The soil basis is assumed rigid. The boundaries are placed far enough from the structure with the use of specific elements and absorbing boundaries ("free field") to reduce the reflection of waves at the edges of the model. The Soil, piles and superstructure mechanical properties are given in Table 1. The piles have an axial rigidity $E_p A_p = 18850$ MN and a flexural rigidity $E_p I_p = 1178$ MN.m^2.

Seismic Loading

The soil-structure system is subjected to two types of loading. The first one corresponds to a harmonic loading with a frequency equal to the soil natural frequency. This loading has very severe consequences and may lead to high internal forces that are not representative to a real earthquake. The second one corresponds to the 1999 Kocaeli earthquake in Turkey with a frequency contents close to the natural frequencies of the soil-structure system.

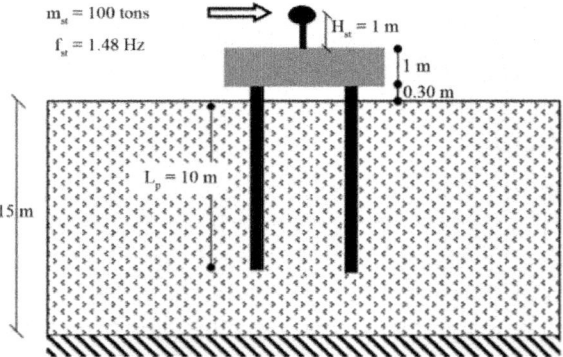

Figure 1: Problem definition.

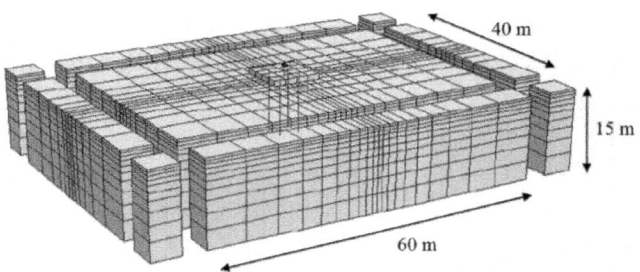

Figure 2: Mesh and boundary conditions.

Table 1: Mechanical properties of soil-pile-superstructure system

	Mechanical properties			
	Density (Kg/m^3)	Young modulus (MPa)	Poisson's ratio	Damping ratio
Soil	1700	8	0.3	$\xi = 5\%$
Pile	2500	24,000	0.3	$\xi = 5\%$
Structure	2500	24,000	0.3	$\xi = 5\%$

In the first step, numerical simulations were performed in the case of a harmonic load of 10 cycles, with a frequency equal to the fundamental frequency of the soil ($f_{loading} = f_{1,sol} = 0.67$ Hz), and an acceleration amplitude of 0.2 g ($V_g = 0.46$ m/s).

In a second step, a real loading is applied as a velocity at the base of the soil mass. The record for the base acceleration, velocity, and displacement waves are shown in Figure 3. It marks a maximum speed of 40 cm/s and a maximum acceleration of 0.247 g. Fourier analysis of the record of the earthquake's velocity results in a power spectrum that reveals a dominant frequency at f = 0.9 Hz (lower peaks are observed at 0.6 and 1.3 Hz) to be compared with the natural frequencies of the soil (0.67 Hz) and the superstructure (1.4 Hz). The Kocaeli earthquake has been chosen because it has frequency contents close to the natural frequencies of the soil-structure system which enhance the development of soil plasticity.

Comparison of Dynamic Forces in the Piles

The forces induced in the piles due to the real seismic loading of Kocaeli, of frequency f = 0.9 Hz, are represented in Table 2 and Figure 4, compared with forces induced in piles due to the harmonic loading of frequency f = 0.67 Hz. Figure 4 shows a significant influence of the dominant frequency loading that can lead to significant efforts values exceeding the bearing capacity of piles and especially when this frequency equals to the natural frequency of the soil. Harmonic loading is very detrimental and causes excessive forces compared to real earthquake loading. So, only real record is used in the next analysis.

Table 2: Piles response for different types of loading

Loading	Acc mass (m/s²)	Acc cap (m/s²)	Max shear force V (KN)	Max. bending moment M (KN.m)
Sinusoidal	34.71	33.1	854.8	3137
Turkey (Kocaeli)	7.36	5.4	145	453.8

EFFECT OF SOIL PLASTICITY

This section deals with the effect of nonlinearities on the behaviour of the soil-pile-structure system, in particular, the influence of soil plasticity on the system response. Numerical simulations are performed using real seismic loading (Turkey, Kocaeli, 1999). The soil behavior is described by an elastic-perfectly plastic Mohr-Coulomb. The case relative to frictional soil is presented in this section.

Plastic Calculation

A parametric study was conducted to know the effect of plasticity for frictional soil on the seismic behaviour of the soil-pile-structure system. The friction angle is considered of 30° and a low cohesion of 2 KPa. To know the influence of dilancy angle, two values were chosen $\psi = 0°$ and $\psi = 20°$. A slight damping of Rayleigh is used for the soil. The behaviour of the cap-structure system is assumed to be elastic.

The extension of plasticity is shown in Figure 5. We can note that the plasticity is localized near the surface due to the low soil confinement at this zone. The seismic loading induces plasticity at the top of soil and the energy is injected into the structure. Plasticization of the soil around the pile head makes it weaker; it leads to the formation of a gap around the pile head which confirms the post seismic observations.

Effect of soil Behavior

Figure 6 shows the internal stresses induced in piles. The variation of the maximum shear force at the top is related to the change of acceleration. For the bending moment, the results at the top are not significantly affected by the change of the dilancy angle. The results are illustrated in Table 3. This parametric study considers two extreme cases of dilancy angles, the intermediate values between 0° and 20° lead to similar tendencies.

EFFECT OF PILES INCLINATION

The effect on the pile inclination on the seismic answer of the system pile-soil-cap is investigated in this section. The case that has been modelled is similar to that earlier, except that the four piles are inclined outwardly. To properly analyze the influence of pile inclination on their

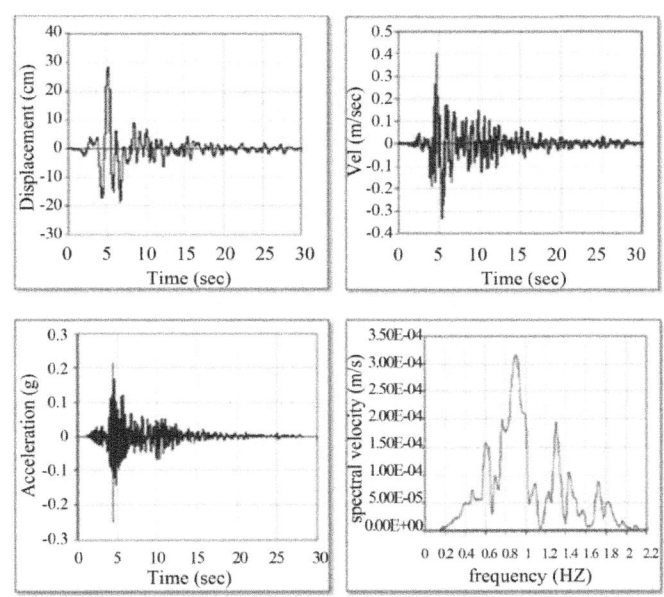

Figure 3: Data of Kocaeli earthquake (Turkey, 1999).

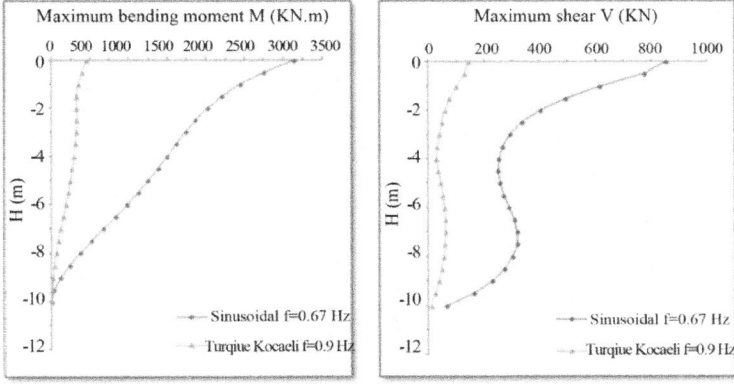

Figure 4: Maximum dynamic forces in the piles.

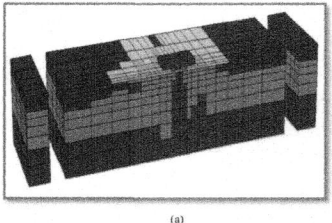 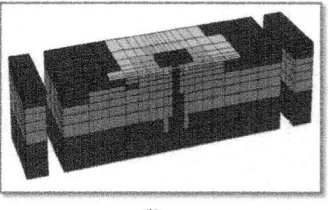

(a)   (b)

Figure 5: Plasticity extension in the case of frictional soil (C = 2 KPa, $\psi = 0°$ and $\psi = 20°$). (a) $\psi = 0°$, (b) $\psi = 20°$.

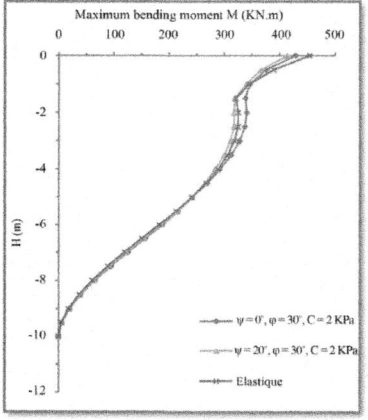 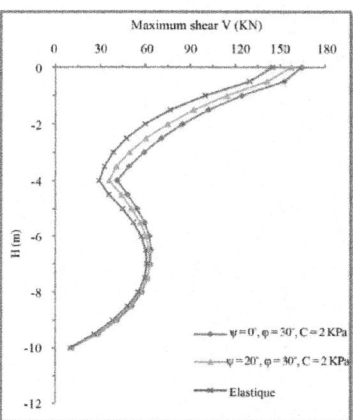

Figure 6: Influence of frictional soil plasticity on the dynamic forces in the piles (Kocaeli earthquake, Turkey, 1999).

Table 3: Influence of frictional soil plasticity on the dynamic forces in the piles

ψ (°)	Acc mass (m/s^2)	Acc cap (m/s^2)	Max shear force V (KN)	Max. bending moment M (KN.m)
Elastic	7.36	5.4	145	453.8
0°	7.431	5.775	164.3	428.5
20°	7.391	6.147	157.7	413.6

seismic response, results of calculations are presented for the two values of inclination angles respectively $\alpha = 10°$ and $\alpha = 20°$. The results are summarized in Figure 7 and Table 4. For the example presented here, the inclination of the pile leads to a reduction of the numerical values of the normal load and lateral displacement of the pile group. However we can remark that along the pile, the values of the moment and the shear have been increased. Table 4 illustrates that on the maximal values of internal forces and the amplification at cap and the structure head have been reduced when the value of the inclination increases except for the value of the bending moment.

CONCLUSIONS

In this paper, we present a three-dimensional numerical modeling of the soil-pile-structure interaction under seismic loading. The effect of the plasticity has been investigated in the case of a frictional soil as well as the effect of the dilancy angle. The numerical modeling has been

Table 4: Influence of inclination on the seismic response of pile groups

Inclination	$\alpha = 0°$	$\alpha = 10°$	$\alpha = 20°$
Amplification at the head of cap	5.40	5.217	3.445
Amplification at the head of structure	7.36	6.526	3.580
Maximum bending moment M (KN.m)	453.8	584.6	657.6
Maximum shear force T (KN)	145	122.8	116.5
Maximum axial force N (KN)	681.1	633.4	446.8

carried by using harmonic excitation and real seismic loading recorded during the Kocaeli earthquake (Turkey, 1999). The effect of the pile inclination has been also analyzed. For simplicity, we consider the case that the piles are embedded in a homogeneous soil. The case of heterogeneous soil could be treated in the future.

The harmonic loading leads to high values of the internal forces (Bending moment, shear) especially when the frequency of the load is near to the proper frequency of the soil. For the example treated here, the plasticity of the soil has a minor effect on the results. For frictional soil, the plasticity spreads from the surface due to the low confinement of the soil in this area. Plasticisation of the soil around the piles head makes them more vulnerable, and the post seismic observations of damaged piles show the formation of a vacuum around the head of the piles.

The inclination of piles leads to a reduction in the lateral amplification of the superstructure resulting from an

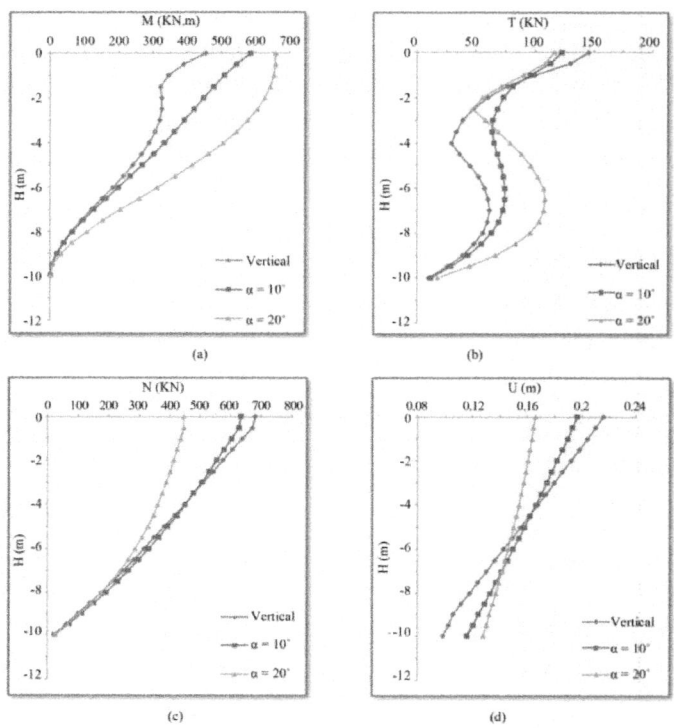

Figure 7: Influence of inclination of piles on the seismic response of pile groups (Registration of Turkey, Vg = 40 cm/s, f = 0.9 Hz, Ag = 0.247 g). (a) Bending Moment, (b) Shear force, (c) Normal force, (d) Displacement.

increase in the rigidity of the system. The inclination of piles can be beneficial for both the dynamic behavior and the behavior of the superstructure. It depends on the interaction of the frequency of the seismic load with the frequencies of the soil-pile-structure. The inclination increases the lateral stiffness of the foundation which, unfortunately, can cause a significant increase in the load transmitted to the foundation of the superstructure. Despite the improved performance of inclined piles, the bending forces at the top of piles are still very significant.

ACKNOWLEDGEMENTS

We thank the Lebanese National Council of Scientific Research for the funding of this work.

REFERENCES

1. M. Sadek and I. Shahrour, "Three-Dimensional Finite Element Analysis of the Seismic Behaviour of Inclined Micropiles," Soil Dynamics and Earthquake Engineering, Vol. 24, 2004, pp. 473-485.

2. M. Sadek and I. Shahrour, "Influence of the Head and Tip Connection on the Seismic Performance of Micropiles," Soil Dynamics and Earthquake Engineering, Vol. 26, No. 6, 2006, pp. 461-468.

3. R. W. Boulanger, C. J. Curras, D. W. Wilson and A. A. Abghari, "Seismic Soil-Pile-Structure Interaction Experiments and Analyses," Journal of Geotechnical and Geoenvironmental Engineering, Vol. 125, No. 9, 1999, pp. 750-759.

4. Y. Chung, "Etude Numérique de L'interaction Sol-PieuStructure sous Chargement Sismique," Thèse de Doctorat, Université de Sciences et Technologie de Lille, 2000.

5. N. Gerolymos, A. Giannakou, I. Anastasopoulos and G. Gazetas, "Evidence of Beneficial Role of Inclined Piles: Observations and Summary of Numerical Analyses," Springer Science and Business Media B. V., 2008.

6. N. Gerolymos, S. Escoffier, G. Gazetas and J. Garnier, "Numerical Modeling of Centrifuge Cyclic Lateral Pile Load Experiments," Earthquake Engineering and Engineering Vibration, Vol. 8, No. 1, 2009, pp. 61-76.

7. Itasca Consulting Group, FLAC, "Fast Lagrangian Analysis of Continua," Vol. I. User's Manual, Vol. II. Verification Problems and Example Applications, 2nd Edition (FLAC3D Version 3.0), Minneapolis, Minnesota, 2005.

8. J. P. Bardet, I. M. Idriss, O'Rourke, N. Adachi, M. Hamada and K. Ishihara, "North America-Japan Workshop on the Geotechnical Aspects of the Kobe," Loma 138 Prieta, and Northridge Earthquake. Report No. 98-36 to National Science Foundation, Air Force Office of Scientific Research, and Japanese Geotechnical Society, Osaka, 1996.

9. K. Tokimatsu, H. Arai and Y. Asak, "Deep Shear Structure and Earthquake Ground Motion. Characteristics in Sumiyoshi Area, Kobe City, Based on Microtremor Measurements," Journal of Structural Engineering (ASCE), Vol. 491, 1997, pp. 37-45.

10. G. Gazetas and G. Mylonakis, "Seismic Soil-Structure Interaction: New Evidence and Emerging Issues. Emerging Issues Paper, Geotechnical Special Publication No 75, ASCE, 2111," Soil Dynamics and Earthquake Engineering, Vol. 26, No. 6, 2006, pp. 461-468.

Chapter 7

DESIGN OF JETTY PILES USING ARTIFICIAL NEURAL NETWORKS

Yongjei Lee[1] Sungchil Lee[2] and Hun-Kyun Bae[3]

[1]Port and Harbor Team, Seoyeong Engineering, Republic of Korea

[2]Department of Computer Design, School of Engineering and Agriculture, Ulaanbaatar University, Mongolia

[3]Department of Global Environment, School of Environment, Keimyung University, 203 Osan Hall, Dalgubul-Daero, Dalsegu, Daegu 1095, Republic of Korea

ABSTRACT

To overcome the complication of jetty pile design process, artificial neural networks (ANN) are adopted. To generate the training samples for training ANN, finite element (FE) analysis was performed 50 times for 50 different design cases. The trained ANN was verified with another FE analysis case and then used as a structural analyzer. The multilayer neural network (MBPNN) with two hidden layers was used for ANN. The framework of MBPNN was defined as the input with the lateral forces on the jetty structure and the type of piles and the output with the stress ratio of the piles. The results from the MBPNN agree well with those from FE analysis. Particularly for more complex modes with hundreds of different design cases, the MBPNN would possibly substitute parametric studies with FE analysis saving design time and cost.

INTRODUCTION

Mooring dolphins are usually constructed when it would be impractical to extend the shore to provide access points to moor vessels. A typical mooring dolphin consists of a platform and several piles supporting the platform, which is so-called jetty. The vertical or battered piles are driven into the seabed. In design practice, deciding whether and where to use the vertical or battered piles is important issue. In the practical design process, the arrangement, the number, and the inclination of the piles are tentatively decided based on previous design experiences and then confirmed through finite element (FE) analysis. Therefore, building and analyzing lots of FE models adopting trial and error process are needed to find the optimum design.

Many researches have been performed to help designers to make decisions. An experimental study showed that the pile group effect is an important factor to resist horizontal loads [1]. For cyclic lateral loading, a zigzag arrangement shows higher resistance than an in-line arrangement. Also it was shown that as the pile center distance increases, the stresses on the front piles decrease, while those on the rear piles increase [2]. When the center distance between piles becomes more than 3~5 times of pile diameter, the group effect decreases so that each pile can be considered as a single pile when the distance reaches 6 times the pile diameter [3]. The battered piles are commonly considered to resist lateral loads solely while the vertical piles resist gravity loads only. This traditional design assumption would make the design process easier but also it usually results in overestimated design. Moreover, it is well known that the vertical piles can also resist bending moments from the lateral loads. Through empirical studies, the p-y method has been proposed and developed by Kondner [4], Reese et al. [5], Scott [6], and Norris [7] to help the design of jetty structure. Though it is still a commonly adopted method, some concepts of the method are based on oversimplified or improper assumptions, especially in the effects of actual soil parameters after pile driving [8].

To overcome the complication of jetty pile design originated from mutual interaction among a number of design parameters, artificial neural networks (ANN) have been introduced in geotechnical engineering [9, 10]. This technique has also been applied successfully in static and dynamic pile systems [11, 12]. Kim et al. [8] predicted the lateral behavior of single and group piles using ANN and compared the results from ANN with the model test results. In this paper, as a suggestive solution of difficulties and cumbersome processes in building and analyzing lots of FE models, the ANN is adopted. Possibility of substituting ANN as a jetty structure analyzer for FE analysis is examined.

METHODOLOGY: APPLICATION OF ANN AS A STRUC-TURE ANALYZER

The jetty design process involves, as mentioned above, searching for the optimum pile pattern which results in the most effective pile usage within feasible design region. The internal forces of the piles of jetty structure subjected to horizontal mooring load vary unexpectedly depending on the inclination of the piles and deployment pattern of piles. Therefore, developing ANN, the input data to ANN are decided as horizontal load exerting on the jetty platform and the information of jetty piles, such as arrangement and inclination of piles, and the output results as the stress ratios of piles to confirm the feasibility of design candidate. Whole concept of methodology adopted in this paper is summarized in Figure 1. As shown in Figure 1, the trained ANN is used as a structural analyzer in this research, placing FE analysis. Firstly, the training samples are generated using FE analysis for various design conditions. It is important that the training samples should be generated from various available design conditions so that the trained ANN may predict adequately when it encounters real new design data. Also the number of training samples should be large enough to avoid overfitting. In this research, total of fifty design cases with different loading conditions and pile patterns are considered for generating training samples through FE analysis.

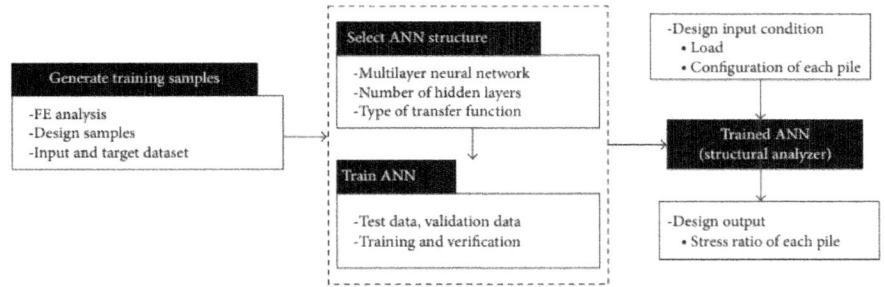

Figure 1: Methodology of design process using ANN.

To construct the ANN architecture with predefined input and output layers, type of ANN, the number of hidden layers and neurons in each hidden layer, and type of transfer function for each layer should be determined. So in this research, because of the complexity of the problem, multilayer back-propagation neural network (MBPNN) with two hidden layers, shown in Figure 2, is adopted to tackle the problem. For the transfer function tangent sigmoid function and pure linear function are adopted for hidden layers and output layer, respectively, since the stress ratio, output from MBPNN, could be either compressive or tensional value. Each neuron of the hidden and output layers has bias and the neurons in one layer are interconnected with the neurons before and after the layer through weights.

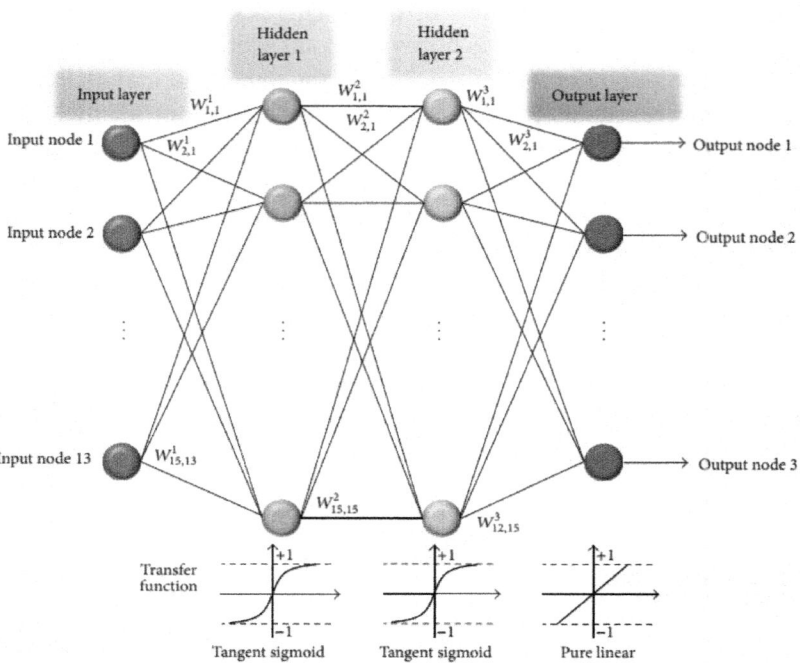

Figure 2: Neural network structure.

Though the number of hidden layers of the MBPNN is determined as two, the performance of the MBPNN will vary depending on the number of neurons in hidden layers. Regarding the number of neurons in each hidden layer, however, there is no general rule to determine. Thus, in this study several neural network architectures with different number of neurons are examined for the best performance and generalization to new data based on the K-fold cross-validation method [13–15]. The performance and generalization of the MBPNN are summarized as an average of root mean squared error (RMSE) from the K-fold cross-validation. After fixing the architecture of MBPNN, training process is conducted to find the optimum values of the biases and weights using all fifty training samples. Finally, the trained MBPNN is used as a structural analyzer to produce the stress ratio of each pile for real design conditions.

DESCRIPTION OF JETTY STRUCTURE FOR ANALYSIS

In this study, a mooring dolphin which was designed for a real project is used. Variations of the pile layout which had been proposed from the early design stage were also considered. The main purpose of the project was to design and build a liquefied natural gas (LNG) terminal at a port area so that the gas product would be transmitted from floating storage and regasification unit (FSRU) to natural gas network onshore by pipelines. The dimension of the platform is 16 m in length, 10 m in width, and 2 m in thickness. The platform is made of reinforced concrete and its piles are made of steel. The plan and elevation view of the testbed mooring dolphin are shown in Figure 3.

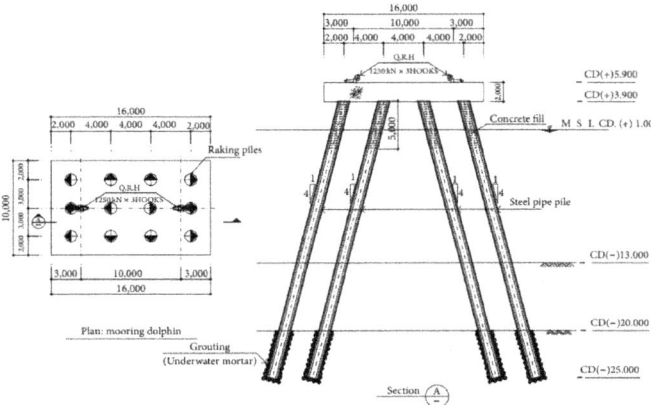

Figure 3: Mooring dolphin for U-project.

Materials

The material properties of C35/45 concrete for the platform are shown in Table 1. In the latest European standard BS EN 206-1 [16], the strength classes are classified using cylinder strength as well as a cube strength. S355 European standard steel is used for most of the structural members [17]. The material properties of S355 steel are shown in Table 2.

Table 1: Material properties of concrete

Unit weight of concrete (dry/submerged)	kN/m^3	23/13.19
Unit weight of RC (dry/submerged)	kN/m^3	24/14.19
Cylinder strength of RC (f_{cu})	N/mm^2	35
Cube strength of RC (f_{cu})	N/mm^2	45
Modulus of elasticity (E_c)	kN/mm^2	29
Poisson's ratio (v)		0.2

Table 2: Material properties of S355 steel.

Thick.	$t \leq 16$	$16 < t \leq 25$	$25 < t \leq 40$	$40 < t \leq 63$	$63 < t \leq 80$	$80 < t \leq 100$	$100 < t \leq 150$
Yield strength (MPa)	355	345	345	335	325	315	295
Tensile strength (MPa)				460–620			
Modulus of elasticity (E)				205 kN/mm^2			
Shear modulus (G)				80 kN/mm^2			
Poisson's ratio (v)				0.3			

Load Conditions

The expected largest FSRU at the mooring dolphin has a capacity of 266,000 m^3 and the largest LNG carrier has a capacity of 177,400 m^3. Maximum mooring force is calculated as 3750 kN. Dead loads are listed in Table 3. All permanent structural members as well as nonstructural members have been considered as dead loads on the structure. Nonstructural member (appurtenance) includes quick release hook (QRH), fender, handrail, and grating. Detail appurtenance loads are shown in Table 4. Pedestrian live load of 4.0 kN/m^2 is assumed. The maximum wave height varies between 2.5 and 2.75 m during a year, but about 60% days of a year wave height is less than 0.5 m. The mean ($T_{m-1,0}$) wave period varies between 2.5 sec and 7 sec. Measurements about 4 km offshore indicate that the typical astronomical velocities are in the order of 0.5 m/s. The largest sea water current speed is 0.7 m/sec at the project site. The wind speed at the location is considered as 18 m/s. The maximum wind speed with 100 years of return period is 32.2 m/sec.

$$\sum_{j\geq 1} \gamma_{G,j} G_{k,j} + \gamma_p P + \gamma_{Q,1} Q_{k,1} + \sum_{j\geq 1} \gamma_{Q,i} \psi_{0,i} Q_k,$$

(1)

Table 3: Dead load

(ton/m^3)	RC	Concrete	Mortar	Steel	Rubble	Fill sand	Sea water
Dry	2.45	2.30	2.15	7.85	0.8	2.0	1.025
Submerged	1.45	1.30	1.15	6.85	1.8	1.0	

Table 4: Appurtenance loads

Nonstructural member	Loads	Remark
Q.R.H	49 kN/EA	
Cone type fender (1800 H (F0.3) or equivalent)	196 kN/EA	Vertical load
Handrail	0.285 kN/m	
Grating	0.838 kN/m^2	

where γ_G, γ_Q are partial factors, P is prestressing, Q is leading variable action, and $\psi 0$ is combination factor.

Soil Conditions

The location of the virtual fixity points was computed by various methods: Chang's method, AASHTO, Hansen's method, and L-pile method. The pile penetration depth under the maximum tensile force was also computed by the Japanese bridge construction standard (2002), API recommended practice 2A-WSD, AASHTO, and Broms' analysis method. Based on those methods, it turned out that the penetration depth of 5 m into the bedrock would provide a fixed boundary condition at the bedrock level.

Design Configuration of Jetty Pile Pattern

Ten different configurations of jetty pile pattern are considered in this study depending on whether vertical or battered, if inclined, the batter direction, and the number of piles (Figure 4). The combination of ten different configurations and five different mooring forces (70%, 80%, 90%, 100%, and 110% of original mooring force) produced 50 FE models and they were analyzed to compute the stress ratios of piles. Among FE models, Patterns 1 and 2 are shown in Figure 5.

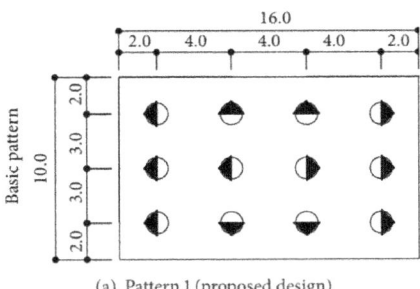

(a) Pattern 1 (proposed design)

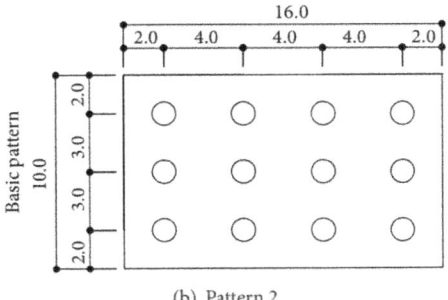

(b) Pattern 2

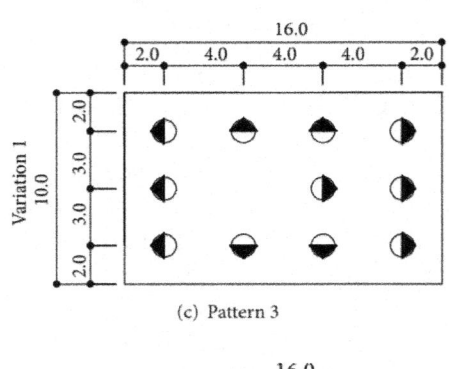

(c) Pattern 3

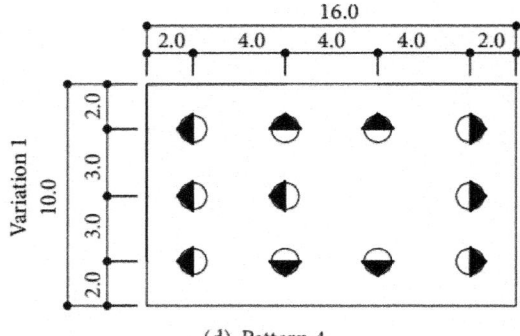

(d) Pattern 4

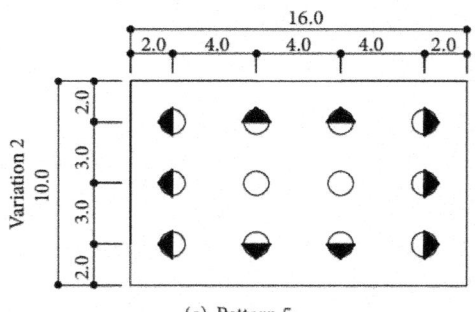

(e) Pattern 5

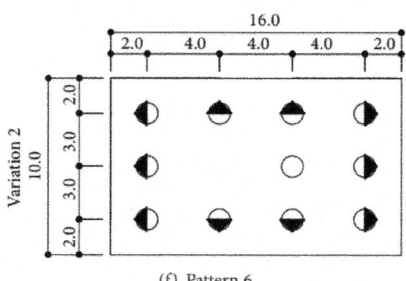

(f) Pattern 6

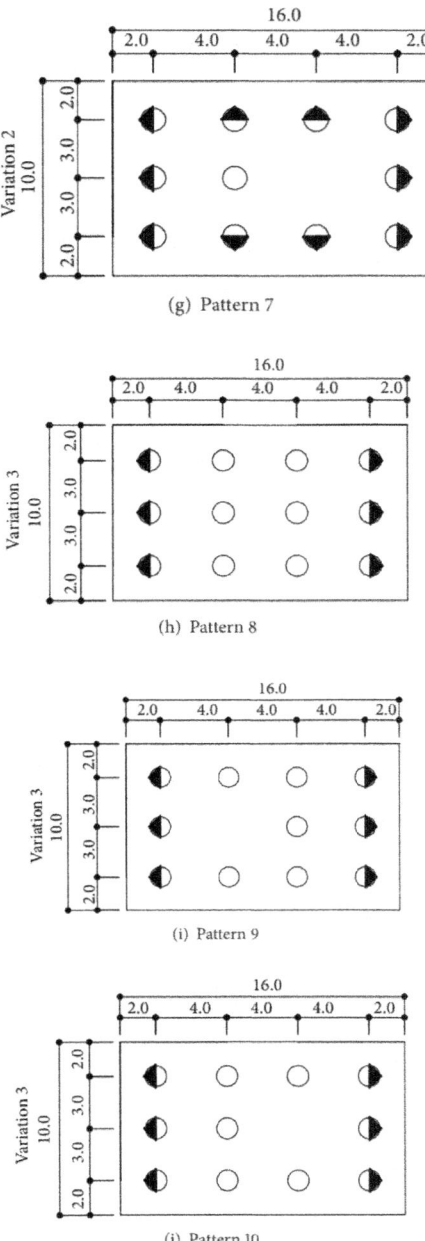

Figure 4: Patterns of piles (circle means vertical pile and circle with triangle means battered pile).

(a)

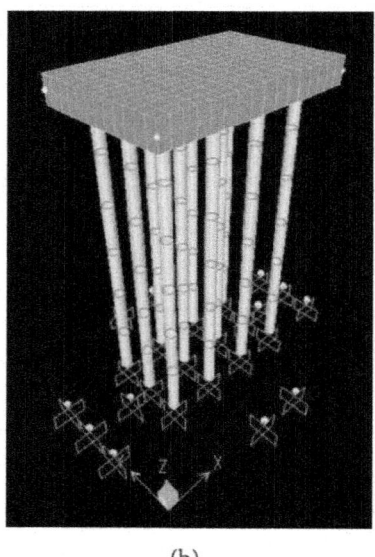

(b)

Figure 5: Finite element modes of Pattern 1 and Pattern 2.

PRELIMINARY FE ANALYSIS

After analyzing 50 FE models of mooring dolphins, the ratios of maximum stress to allowable stress of each pile were obtained under given loading condition as shown in Figure 6. Here, a pile pattern with the less number of required piles as well as with the smaller stress ratio is considered as the improved.

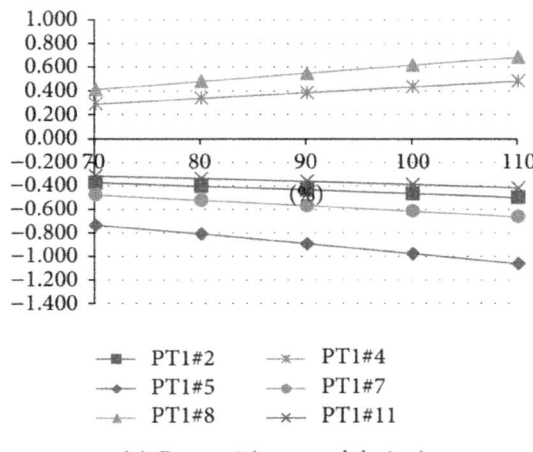

(a) Pattern 1 (proposed design)

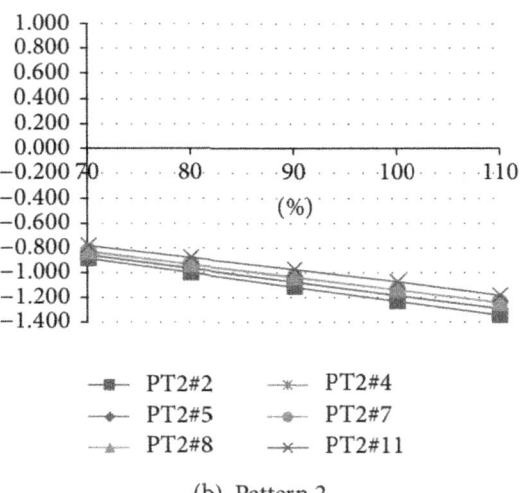

(b) Pattern 2

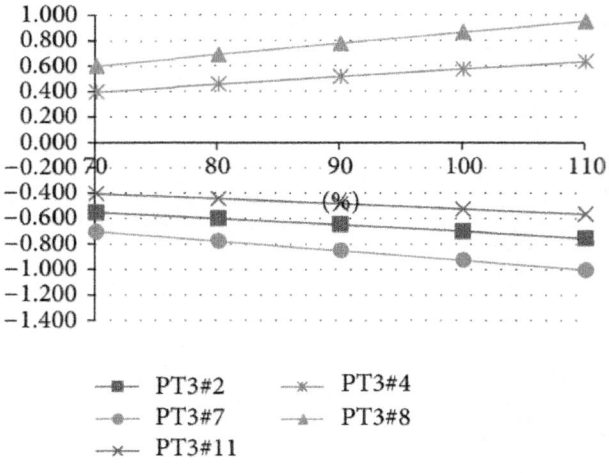

(c) Pattern 3

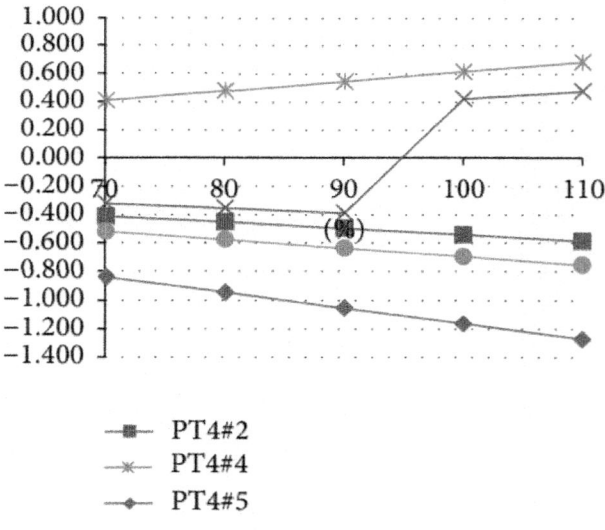

(d) Pattern 4

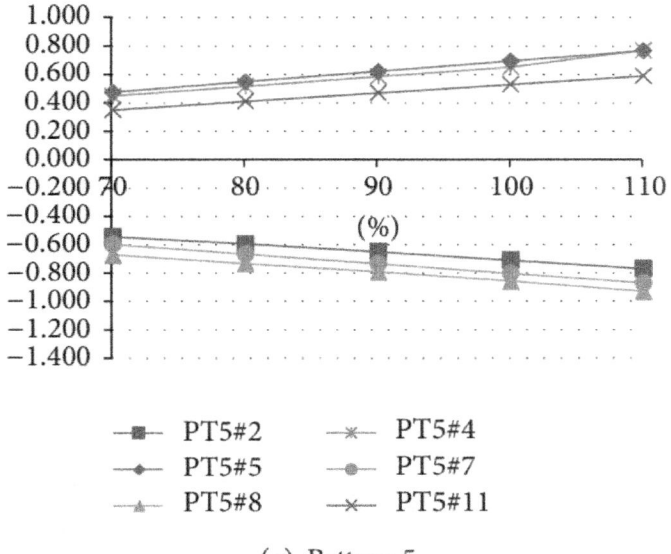

(e) Pattern 5

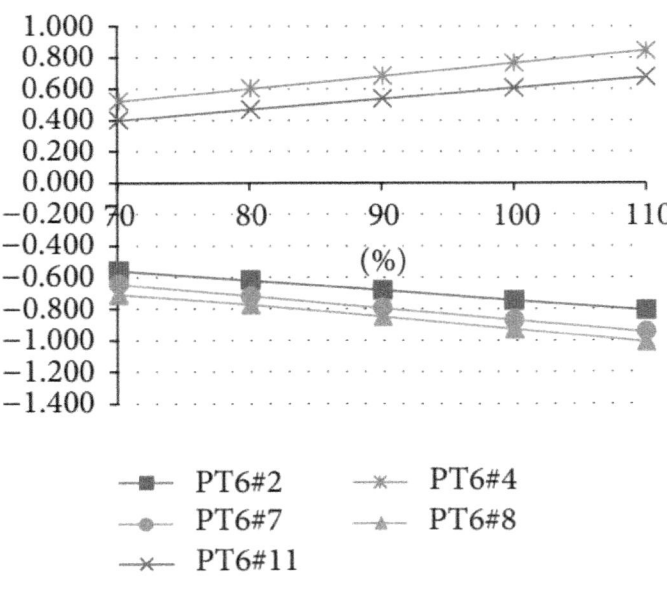

(f) Pattern 6

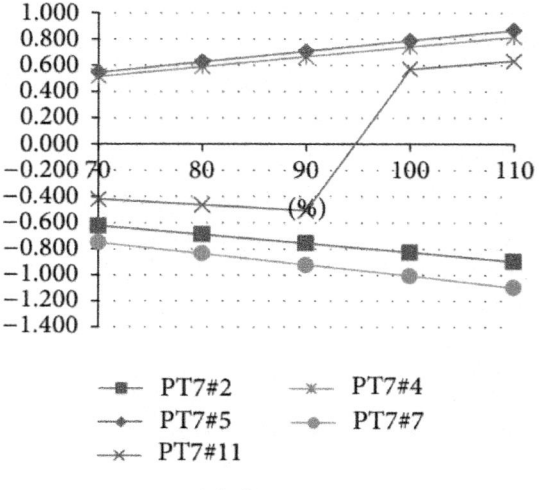

▪ PT7#2	✳ PT7#4
◆ PT7#5	● PT7#7
✕ PT7#11	

(g) Pattern 7

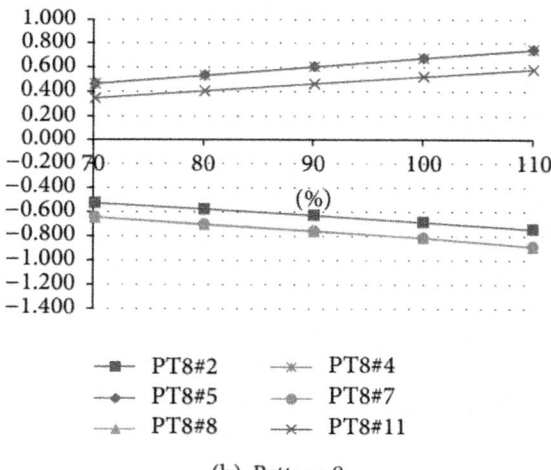

▪ PT8#2	✳ PT8#4
◆ PT8#5	● PT8#7
▲ PT8#8	✕ PT8#11

(h) Pattern 8

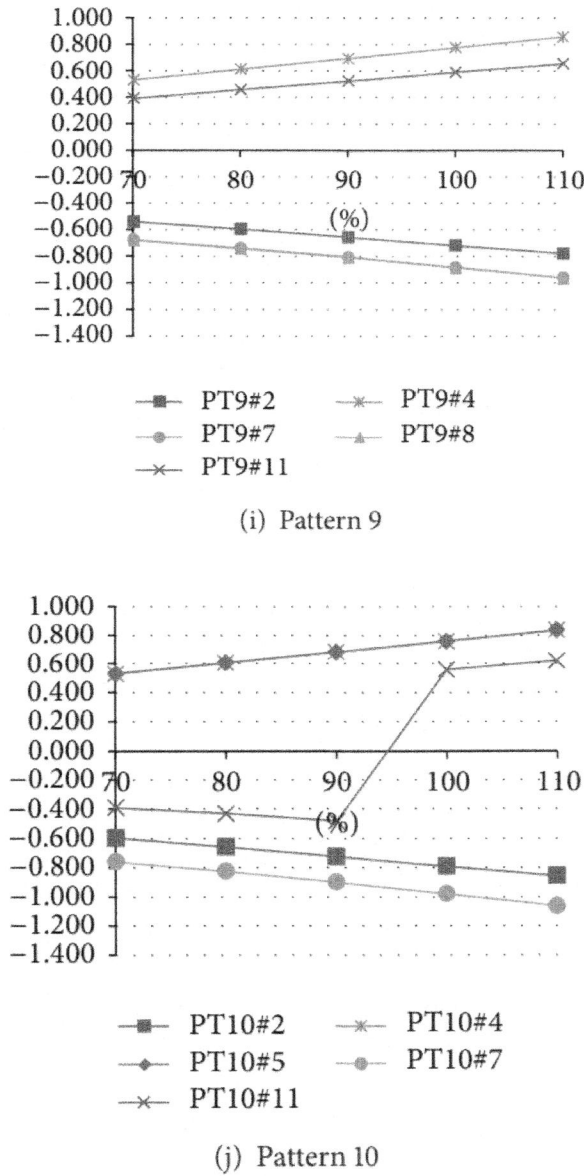

(i) Pattern 9

(j) Pattern 10

Figure 6: Maximum stress ratio of each pile.

Comparison of Battered and Vertical Piles

In Pattern 1, the absolute values of the stress ratio are all less than unity except Pile 5. Pile 5 shows the maximum compressive stress and Pile 8 shows the maximum tensile stress. Pattern 2, whose piles are all vertical, shows compressive stress and most of the stress ratios are greater than unity.

Patterns 3 and 4

When Pile 5 is absent (Pattern 3), the tensile stress of Pile 8 becomes bigger than that of the proposed design (Pattern 1). In Pattern 3, the compressive force on Pile 5 in Pattern 1 redistributes to the adjacent piles. Considering all stress ratios and the number of piles, Pattern 3 can be considered as more improved design than the proposed one. Pile 11 of Pattern 4 is compressive within 90% of applied force but it turns to be tensile when the mooring force is equal to or more than 100%. In this case, the absence of Pile 8 causes a rapid stress change in Pile 11 and design of reinforcing bars in concrete platform is difficult; therefore Pattern 4 shall be avoided.

Patterns 5, 6, and 7

When it is compared to Pattern 1, Pile 4 of Pattern 5 remains tensile. However, Piles 4 and 8 of Pattern 1 change to be compressive in Pattern 5, while Piles 5 and 11 become tensile. The use of vertical piles in Pattern 5 makes stress sign changed. All absolute values of stress ratio of Pattern 5 are less than unity. In this point of view, Pattern 5 is more effective than the proposed design. With absence of Pile 5, Pattern 6 shows slightly higher stress than Pattern 5. Similar to Pattern 4, the absence of Pile 8 causes rapid stress change in Pile 11 of Pattern 7. Pattern 7 shows higher compressive stress than Pattern 5.

Patterns 8, 9, and 10

The stress ratios of Patterns 8, 9, and 10 are similar to those of Patterns 5, 6, and 7 but they are not compatible for horizontal load direction change.

ARCHITECTURE AND TRAINING OF MBPNN

Training Samples

The first neuron of the input layer is assigned for load condition, and the other input neurons take the information of piles. To distinguish the battered and the vertical piles, the numbers "1" and "2" are assigned as neuron input values for each pile. To the location where pile is absent, the number "0" is assigned to the corresponding input neuron. For the value of mooring force corresponding to the value of input neuron 1 which is very big compared with the values of the other input neurons might lead to a failure of MBPNN training, the mooring force is normalized to the mooring force of 3750 kN. Table 5 shows the example of input values and corresponding pile locations.

Table 5: Example of neuron inputs for neural network training process

Input neuron		Remark
Number	Assigned value	
1	1.0	Normalized value of horizontal mooring force to 3750 kN
2	1.0	Pile 1: battered
3	1.0	Pile 2: battered
⋮	⋮	⋮
12	2.0	Pile 11: vertical
13	2.0	Pile 12: vertical

In this study, MBPNN, with two hidden layers, utilizing back-propagation process is used. To obtain the training samples, five load cases—70%, 80%, 90%, 100%, and 110% of mooring force of 3750 kN—are applied to 10 different pile patterns of jetty structures. The combination of 10 jetty pile patterns and 5 load cases makes 50 training cases in total. Table 6 shows an example of input and target output of training samples.

Table 6: Example of training samples

(a)

	Input neuron (load and information of each pile)								
Cases	1	2	3	4	5	...	11	12	13
	(Load)	(Pile 1)	(Pile 2)	(Pile 3)	(Pile 4)		(Pile 10)	(Pile 11)	(Pile 12)
1	0.7	1	1	1	1	...	1	1	1
2	0.7	2	2	2	2	...	2	2	2
3	0.7	1	1	1	1	...	1	1	1
⋮	⋮	⋮	⋮	⋮	⋮	⋮	⋮	⋮	⋮
48	1.1	1	1	1	2	...	1	1	1
49	1.1	1	1	2	0	...	1	1	1
50	1.1	1	1	1	2	...	1	1	1

(b)

	Output neuron (stress ratio of each pile)							
Cases	1	2	3	4	5	...	11	12
	(Pile 1)	(Pile 2)	(Pile 3)	(Pile 4)	(Pile 5)		(Pile 11)	(Pile 12)
1	−0.364	−0.363	−0.364	0.302	−0.727	...	−0.307	−0.305
2	−0.877	−0.877	−0.877	−0.847	−0.847	...	−0.771	−0.771
3	−0.547	−0.548	−0.547	0.404	0.404	...	−0.399	−0.397
⋮	⋮	⋮	⋮	⋮	⋮	⋮	⋮	⋮
48	−0.750	−0.751	−0.750	0.737	0.737	...	0.573	0.574
49	−0.788	−0.787	−0.788	0.856	0.856	...	0.651	0.651
50	−0.863	−0.864	−0.863	0.831	0.830	...	0.613	0.616

Construction of MBPNN Architecture and Training

Since the performance and generalization of MBPNN to new design data will vary depending on the number of neurons in hidden layers, four different topologies of MBPNN with different number of neurons in hidden layers are examined: (1) 13 (input neurons)-15(1st hidden layer)-15 (2nd hidden layer)-12 (output layers),(2) 13-10-10-12, (3) 13-7-15-12, and (4) 13-15-10-12. The number of neurons in hidden layers of the first MBPNN model (13-15-15-12) is greater than that of input or output layers and vice versa in the second model (13-10-10-12). In this paper, the K-fold cross-validation method is used to assess the generalization of model and to select the best architecture of MBPNN. For the K-fold cross-validation the fifty training samples are randomly divided into 10 subsets, that is, 10-fold cross-validation. In the K-fold cross-validation one subset is assigned as validation data set and the other nine subsets as training data set. Each MBPNN model is trained using the training data set and RMSE is computed for the validation data set. This procedure continues 10 times changing validation data set and training data set. Finally, a MBPNN model with the least averaged RMSE is selected as the best model.

The neural network toolbox provided by commercial program MATLAB was used to construct and for training of MBPNN models. The Levenberg-Marquardt method with back-propagation process was adopted for the optimization algorithm for training. The Levenberg-Marquardt method is considered to be effective for the complicated MBPNN for the fastest training [19, 20]. The objective function in the neural network training is defined as the minimization of mean squared error as shown:

$$\text{Objective Function} = \text{Minimization}: \frac{1}{n}\sum_{i=1}^{n}(y_i - f(x_i,\delta))^2, \tag{2}$$

where n = total number of training pattern, y_i = ith target, x_i = ith input, and δ = weights and biases of neural network.

In the Levenberg-Marquardt method, the optimum weights and biases are searched using

$$\delta_{k+1} = \delta_k - \left(J^T J + \mu\,\text{diag}\left(J^T J\right)\right)^{-1} J^T e, \tag{3}$$

where δ_{k+1} = $(k+1)$th weights and biases, J = Jacobian matrix, μ = adaptive value, and e = error.

The successful performance of the Levenberg-Marquardt method depends on the choice of μ. Where the gradient is small, the search movement should be large so that the slow convergence is avoided. However, it should be small for the steeper gradient region. The initial value of μ was assumed as 0.001, and the increase and decrease factor of μ were assigned as 10 and 0.1, respectively. Thus, during the training, μ will take the value of 0.001×(increase factor or decrease factor) n, where n is zero or natural number. The training process starts with the initial value and at the second step the objective function (i.e., mean squared error) is computed with the previous value of $\mu = 0.001$ and $\mu = 0.001 \times$ (decrease factor). If both of these values do not result in good performance, a new value of $\mu = 0.001 \times$ (increase factor) is adopted for the next step. In the following steps, the best μ value is searched among "previous μ value," "(previous μ value)×(decrease factor)," and "(previous μ value)×(increase factor) n ." If a previous μ value results in reduction of the objective function, the value is not changed.

Table 7 summarizes the averaged RMSE of each MBPNN model from the K-fold cross-validation. Interestingly the first model (13-15-15-12) which is the most complex one among the four models does not show the least averaged RMSE, rather than the second model of which the number of neurons in hidden layers is between that of input and output layer which shows the least. From the K-fold cross-validation results, the second MBPNN model with 10 neurons for each hidden layer is selected as the best architecture.

Table 7: RMSE of each MBPNN model from K-fold cross-validation

Round	MBPNN model			
	13-15-15-12	13-10-10-12	13-7-15-12	13-15-10-12
1	0.0130	0.0986	0.0482	0.1205
2	0.0606	0.0125	0.0796	0.0118
3	0.0192	0.0086	0.1495	0.0238
4	0.0295	0.0046	0.0675	0.0264
5	0.1138	0.0986	0.1181	0.0078
6	0.1733	0.0383	0.1607	0.0413
7	0.0147	0.0346	0.1218	0.1053
8	0.1147	0.0643	0.0791	0.0730
9	0.1396	0.0461	0.1098	0.2451
10	0.0052	0.0258	0.0343	0.0131
Averaged	0.0684	0.0432	0.0969	0.0668

After fixing the architecture of the MBPNN, the MBPNN was trained again with all the training samples. The training process was completed at 147 epochs, as shown in Figure 7, and terminated at performance goal of 10−6 which was set as one of the termination conditions. Figure 8 shows the gradient changes of the problem surface and changes of the μ value during the training process. If a μ value gives the reduction of objective function, the value is kept for the following steps. However, if it does not result in good performance, it is updated using the increase or decrease factor. To find the optimum point, the μ value is continuously updated during the training process and with the change of the μ value the searching direction and gradient of problem surface are changed. As long as the gradient is large enough to improve the training performance, the μ value is unchanged; however, when the gradient gets smaller and the training performance gets worse, which means the surface of the objective function becomes flat, the μ value is updated. Comparing the training performance graph and the μ graph, it is observed that the μ value starts from the initial value of 0.001 and is updated at epochs 2, 7, 16, 42, and 82. In Figure 8 with the constant μ values the performance and gradient become flat; however, the training performance and gradient are improved dramatically at those epochs.

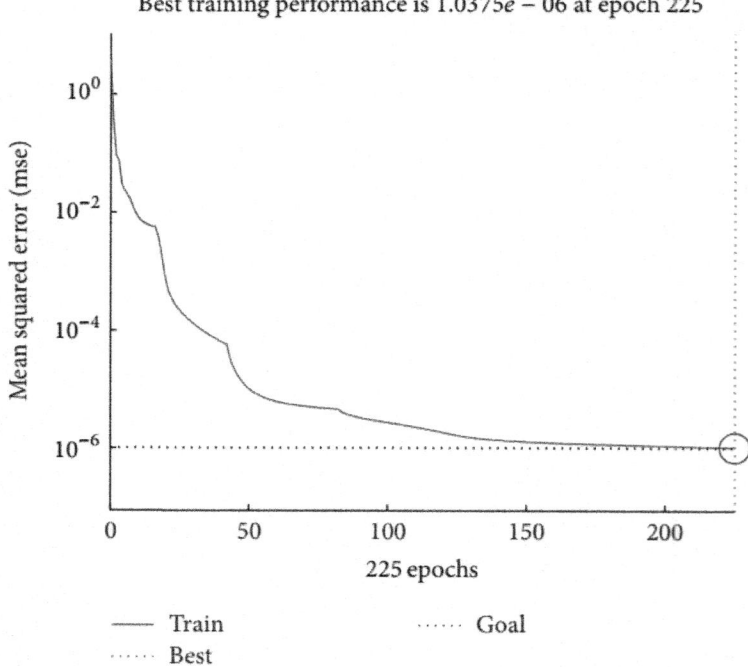

Best training performance is 1.0375e − 06 at epoch 225

Figure 7: Training performance.

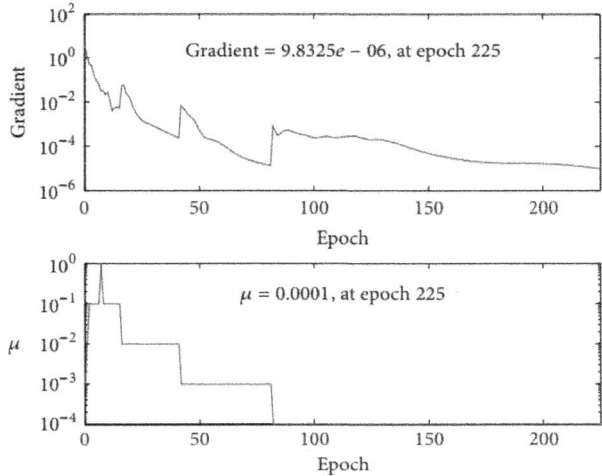

Figure 8: Change of gradient and μ at each training epoch.

Verification of Training

Cases 3, 15, 27, 39, and 50 were selected for verification of MBPNN training. Table 8 shows the comparison of 5 known targets of training cases and the simulated results from the trained MBPNN. The values in the last row are root mean squared error (RMSE) between the known targets and the simulated values. At the locations of no pile, the target value is 0 and the simulation results also show similar value and RMSE is very small. This ensures that the MBPNN is well trained.

Table 8: Simulation results

Pile number	\multicolumn{10}{c}{Number of training cases}
	\multicolumn{2}{c}{3}	\multicolumn{2}{c}{15}	\multicolumn{2}{c}{27}	\multicolumn{2}{c}{39}	\multicolumn{2}{c}{50}
	Target	Simul.	Target	Simul.	Target	Simul.	Target	Simul.	Target	Simul.
1	−0.6050	−0.6051	−0.6030	−0.6035	−0.7580	−0.7580	−0.7120	−0.7117	−0.7500	−0.7492
2	−0.6050	−0.6055	−0.6030	−0.6032	−0.7580	−0.7588	−0.7120	−0.7121	−0.7510	−0.7496
3	−0.6050	−0.6051	−0.6030	−0.6035	−0.7580	−0.7580	−0.7120	−0.7117	−0.7500	−0.7492
4	0.5230	0.5227	0.6090	0.6082	0.6680	0.6673	0.6540	0.6537	0.6400	0.6398
5	0.5220	0.5227	$0^\dagger$	0.0001	0.7080	0.7092	0.6940	0.6937	$0^\dagger$	−0.0004
6	0.5230	0.5227	0.6090	0.6082	0.6680	0.6672	0.6540	0.6538	0.6400	0.6399
7	−0.7660	−0.7649	−0.7440	−0.7433	−0.9210	−0.9210	−0.8060	−0.8057	−0.9990	−0.9990
8	$0^\dagger$	−0.0008	−0.7490	−0.7475	$0^\dagger$	−0.0004	−0.8600	−0.8599	0.9600	0.9593
9	−0.7660	−0.7657	−0.7440	−0.7432	−0.9210	−0.9203	−0.8060	−0.8059	−0.9990	−1.0010
10	−0.3970	−0.3963	0.4530	0.4528	−0.5100	−0.5096	0.5310	0.5302	−0.5580	−0.5580
11	−0.3980	−0.3974	0.4520	0.4521	−0.5080	−0.5093	0.5300	0.5294	−0.5600	−0.5600
12	−0.3970	−0.3963	0.4530	0.4528	−0.5100	−0.5096	0.5310	0.5302	−0.5580	−0.5580
RMSE	\multicolumn{2}{c}{0.0008}	\multicolumn{2}{c}{0.0004}	\multicolumn{2}{c}{0.0007}	\multicolumn{2}{c}{0.0006}	\multicolumn{2}{c}{0.0005}

RMSE: root mean squared error.
†No pile at this location.

Design with Trained MBPNN

Through the trained MBPNN, the stress ratios of jetty piles were obtained under different loading conditions which were not included in the training samples. The feasibility of the MBPNN was verified by comparing the results from FE model and the MBPNN. Table 9 shows the stress ratios computed by FE analysis and the MBPNN. The results from the MBPNN are very close to the FE analysis results. The RMSE is also very small regardless of the pile patterns.

Table 9: Comparison of stress ratios obtained by MBPNN and FE analysis

| Pile number | Patterns (75% of original mooring force) | | | | | | | | | |
| | 1 | | 2 | | 3 | | 4 | | 5 | |
	Anlys.	ANN	Anlys.	ANN	Anlys.	ANN	Anlys.	ANN	Anlys.	ANN
1	−0.3800	−0.3803	−0.9350	−0.9333	−0.5710	−0.5697	−0.4290	−0.4297	−0.5720	−0.5710
2	−0.3790	−0.3788	−0.9350	−0.9335	−0.5710	−0.5701	−0.4270	−0.4281	−0.5730	−0.5713
3	−0.3800	−0.3803	−0.9350	−0.9333	−0.5710	−0.5697	−0.4290	−0.4297	−0.5720	−0.5710
4	0.3260	0.3259	−0.9020	−0.9004	0.4330	0.4333	0.4500	0.4486	0.4830	0.4834
5	−0.7610	−0.7633	−0.9020	−0.9012	0†	−0.0004	−0.8870	−0.8873	0.5120	0.5127
6	0.3260	0.3261	−0.9020	−0.9004	0.4330	0.4333	0.4500	0.4487	0.4830	0.4834
7	−0.4910	−0.4885	−0.8690	−0.8679	−0.7360	−0.7347	−0.5450	−0.5442	−0.6350	−0.6357
8	0.4580	0.4580	−0.8740	−0.8727	0.6530	0.6529	0†	0.0005	−0.7050	−0.7038
9	−0.4910	−0.4790	−0.8690	−0.8678	−0.7360	−0.7357	−0.5450	−0.5437	−0.6350	−0.6346
10	−0.3140	−0.3154	−0.8190	−0.8147	−0.4170	−0.4168	−0.3320	−0.3308	0.3820	0.3825
11	−0.3160	−0.3171	−0.8190	−0.8146	−0.4190	−0.4187	−0.3320	−0.3313	0.3810	0.3814
12	−0.3140	−0.3154	−0.8190	−0.8147	−0.4170	−0.4168	−0.3320	−0.3308	0.3820	0.3825
RMSE	0.0037		0.0025		0.0007		0.0010		0.0008	

| Pile number | Patterns (75% of original mooring force) | | | | | | | | | |
| | 6 | | 7 | | 8 | | 9 | | 10 | |
	Anlys.	ANN	Anlys.	ANN	Anlys.	ANN	Anlys.	ANN	Anlys.	ANN
1	−0.5900	−0.5918	−0.6550	−0.6556	−0.5590	−0.5587	−0.5750	−0.5749	−0.6330	−0.6336
2	−0.5900	−0.5912	−0.6240	−0.6563	−0.5600	−0.5593	−0.5740	−0.5747	−0.6340	−0.6341
3	−0.5900	−0.5918	−0.6550	−0.6556	−0.5590	−0.5587	−0.5750	−0.5749	−0.6330	−0.6336
4	0.5610	0.5598	0.5540	0.5540	0.4920	0.4920	0.5680	0.5670	0.5620	0.5615
5	0†	−0.0001	0.5880	0.5869	0.4920	0.4921	0†	−0.0001	0.5610	0.5599
6	0.5610	0.5599	0.5540	0.5540	0.4920	0.4920	0.5680	0.5670	0.5620	0.5615
7	−0.6840	−0.6834	−0.7930	−0.7928	−0.6770	−0.6757	−0.7130	−0.7107	−0.7990	−0.7973
8	−0.7430	−0.7420	0†	−0.0003	−0.6830	−0.6825	−0.7180	−0.7158	0†	0.0003
9	−0.6840	−0.6833	−0.7930	−0.7917	−0.6770	−0.6750	−0.7130	−0.7106	−0.7990	−0.7981
10	0.4350	0.4353	−0.4410	−0.4396	0.3710	0.3718	0.4200	0.4200	−0.4170	−0.4161
11	0.4350	0.4344	−0.4420	−0.4398	0.3700	0.3705	0.4190	0.4192	−0.4180	−0.4169
12	0.4350	0.4353	−0.4410	−0.4396	0.3710	0.3718	0.4200	0.4200	−0.4170	−0.4161
RMSE	0.0010		0.0094		0.0008		0.0013		0.0009	

RMSE: root mean squared error.
†No pile at this location.

CONCLUSIONS

In this paper, the application of MBPNN as a structural analyzer for jetty structures is explored. The framework of MBPNN is defined as the input with the lateral forces on the jetty structure and the type of piles and the output with the stress ratios of the piles. For the highly complex jetty pile patterns the results from the MBPNN show very good agreement with those from FE analysis. With the more training samples and the expansion of input parameters for jetty structure design, the MBPNN shows possibility to replace the repetitive and time-consuming FE analysis. Although only 50 cases have been modeled for study purpose in this paper, the merit of MBPNN would be clearer as the number of cases increases.

REFERENCES

1. J. S. Bae, T. Y. Kim, and S. H. Kim, "The lateral behavior characteristics of group piles in sand ground,"Journal of Korea Society of Civil Engineers C, vol. 20, no. 6, pp. 517–524, 2000.

2. J. S. Bae and S. H. Kim, "The behavior of group piles for pile arrangement under repeated lateral loads,"Journal of Korea Society of Civil Engineers, vol. 23, no. 4, pp. 231–239, 2003.

3. J. Y. Paek, J. Y. Cho, S. S. Jeong, and T. J. Hwang, "Shaft group efficiency of friction pile groups in deep soft clay," Journal of Korea Society of Civil Engineers, vol. 32, no. 2, pp. 49–60, 2012.

4. R. L. Kondner, "Hyperbolic stress-strain response: cohesive soils," Journal of the Soil Mechanics and Foundations Division, vol. 89, no. 1, pp. 115–143, 1963.

5. L. C. Reese, W. R. Cox, and F. D. Koop, "Analysis of laterally loaded piles in sand," in Proceedings of the 6th Offshore Technology Conference, pp. 473–483, Dallas, Tex, USA, 1974.

6. R. F. Scott, "Analysis of centrifuge pile tests: simulation of pile driving," Research Report OSAPR Project 13, American Petroleum Institute, 1980.

7. G. Norris, "Theoretically based BEF laterally loaded pile analysis," in Proceedings of the 3rd International Conference on Numerical Methods in Offshore Piling, pp. 361–386, Navtes, 1986.

8. B. T. Kim, S. Kim, and S. H. Lee, "Prediction of lateral behavior of single and group piles using artificial neural networks," KSCE Journal of Civil Engineering, vol. 5, no. 2, pp. 185–198, 2001.

9. W. T. Chan, Y. K. Chow, and L. F. Liu, "Neural network: an alternative to pile driving formulas,"Computers and Geotechnics, vol. 17, no. 2, pp. 135–156, 1995. ··

10. G. W. Willis, C. Yao, R. Zhao, and D. Penumadu, "Stress-strain modeling of sands using artificial neural networks," Journal of Geotechnical Engineering, vol. 121, no. 5, pp. 429–435, 1995. ··

11. I. Lee and J. Lee, "Prediction of pile bearing capacity using artificial neural networks," Computers and Geotechnics, vol. 18, no. 3, pp. 189–

200, 1996.

12. C. I. Teh, K. S. Wong, A. T. C. Goh, and S. Jaritngam, "Prediction of pile capacity using neural networks,"Journal of Computing in Civil Engineering, vol. 11, no. 2, pp. 129–138, 1997.

13. S. Arlot and A. Celisse, "A survey of cross-validation procedures for model selection," Statistics Surveys, vol. 4, pp. 40–79, 2010.

14. T. Fushiki, "Estimation of prediction error by using K-fold cross-validation," Statistics and Computing, vol. 21, no. 2, pp. 137–146, 2011.

15. R. Hornung, C. Bernau, C. Truntzer, T. Stadler, and A. Boulesteix, "Full versus incomplete cross-validation: measuring the impact of imperfect separation between training and test sets in prediction error estimation," Tech. Rep. 159, Department of Statistics, University of Munich, 2014.

16. CEN/TC 104, Concrete—Part 1: Specification, performance, production and conformity, EN 206-1, 2000.

17. CEN/TC 104, "Structural steel part 1: general technical delivery conditions," Tech. Rep. EN 10025-1, 2004.

18. British Standard Institution, "Maritime works—part 2: Code of practice for the design of quay walls, jetties and dolphins," BS 6349-2, 2010.

19. K. Levenberg, "A method for the solution of certain non-linear problems in least squares," Quarterly of Applied Mathematics, vol. 2, pp. 164–168, 1944. · ·

20. D. W. Marquardt, "An algorithm for least-squares estimation of nonlinear parameters," SIAM Journal on Applied Mathematics, vol. 11, no. 2, pp. 431–441, 1963. ·

Chapter 8

WIND LOAD ANALYSIS OF TALL CHIMNEYS WITH PILED RAFT FOUNDATION CONSIDERING THE FLEXIBILITY OF SOIL

B. R. Jayalekshmi, S. V. Jisha , R. Shivashankar

National Institute of Technology Karnataka, Surathkal, Karnataka, India

ABSTRACT

Soil–structure interaction (SSI) analysis was carried out for tall reinforced concrete chimneys with piled raft foundation subjected to wind loads. To understand the significance of SSI, four types of soil were considered based on different material properties. Chimneys of different elevations and different ratios of height to base diameter of chimney were selected for the parametric study. The thickness of raft of piled raft foundation was also varied based on different ratios of outer diameter to thickness of raft. The chimneys were assumed to be located in open terrain and subjected to a maximum wind speed of 50m/s. The along-wind and across-wind loads were computed according to IS: 4998 (Part 1)-1992 and applied along the height of the chimney. The analysis was carried out using three-dimensional finite element technique based on the direct method of SSI. The linear elastic material behaviour was assumed for the integrated chimney–foundation–soil system. The radial and tangential moments, lateral deflection and base moment of chimney were evaluated through SSI analysis and compared with the response obtained from chimney with fixed base. The base moment of chimney considerably reduces due to the effect of SSI. It is found that the variation of different responses in chimney due to the effect of SSI depends significantly on the geometrical properties of chimney and foundations. The response variation at base for a distance of 1/40th of the height of chimney should be considered for a safe design.

INTRODUCTION

Chimneys are very important structures in any industry and are used to discharge the pollutants to higher atmosphere. The chimney elevations have gone up progressively from 100m to more than 400m due to the high demand of pollution control. Chimneys have unique geometrical features of slender dimensions and tapering geometry and, therefore, the analysis and design of such kind of structure should be treated separately from other forms of tower structures. These tall chimneys are very sensitive to wind loads.

The dynamic wind effects on chimney are predicted by analytical procedures but they are somewhat complicated, time-consuming and require specialized software. If the modes of chimney are well separated then simplified design procedures can be used. The simplified design techniques such as static or quasi-static methods that account for the wind effect of chimney were given by Manohar (1985). The effect of wind on such tall freestanding structures has two components, namely along-wind and across-wind load conditions. The chimney is subjected to gust buffeting in the along-wind direction (due to drag forces), and also to possible vortex shedding in the across-wind direction. The along-wind and across-wind loads can be estimated using most of the design codes for chimneys (CICIND2001; ACI 307-082008; Koten2005; IS: 4998(Part 1)-19922003). The gust factor method (Davenport1967) is one of the prominent methods widely used for the along-wind load calculation and various modifications have been made on gust factor method by many researchers (Simiu1976; Solari1982). The international codal recommendations were reviewed by Menon and Rao (1997a,b) to determine the design moments for along-wind and across-wind load conditions in reinforced concrete chimneys. Different expressions were formulated by several researchers (Vickery and Clark1972; Kwok and Melbourne1981; Davenport1995; Melbourne1997) to evaluate the response of structures due to across-wind load conditions. An empirical method was presented by Arunachalam et al. (2001) for correlating the rms lift coefficient due to vortex shedding and Strouhal number. The above studies neglect the effect of foundation and underlying soil.

The annular raft foundations are more reasonable and economical than the full circular raft for industrial chimneys. If the geotechnical conditions are not favourable for raft foundations, piled foundations can also be used. Skin friction piles are more suitable to chimney foundations than end bearing piles, since greater uplift capacity is generally available (Turner2005). It is seen that generally the analysis of these foundations is carried out without considering the effect of super structure (Chu and Afandi1966; Brown1969; Melerski1990; IS: 11089-19842002; Dewaikar and Patil2006). Chaudhary (2007) investigated the effectiveness of pile foundation in reducing settlement by comparing the results of piled raft foundation and the raft foundation alone. Analysis of foundation without structure and analysis of structure without foundation may give incorrect results in the different responses of structure and foundation.

Studies by Pour and Chowdhury (2008) proved that base moment of tall chimney founded on soft soil increase up to 10% due to along-wind load and decrease up to 50% due to across-wind load that may affect the design forces. The effect of long-duration earthquakes as well as the higher mode participation in a 215-m tall chimney considering SSI is studied by Mehta and Gandhi (2008). Considerable reduction in the bending moments in the annular raft foundation of tall chimneys due to the effect of flexibility of supporting soil under along-wind load is reported (Jayalekshmi et al.2011). The above studies point towards the need of further investigation on the SSI analysis of chimneys with piled raft foundations.

SOIL–STRUCTURE INTERACTION

The response of the structure affects the motion of supporting soil and the movement of supporting soil influences the structural behaviour. This inter-dependency of response between the structure and the soil is referred as SSI. Depending on the modelling method for the soil stratum, all the SSI problems can be classified into two main categories, namely direct method and substructure method (Wolf1985). The direct method evaluates the response of structure and its surrounding soil in a single analysis step by subjecting the combined soil–structure system to applied loads. The finite element analysis can be easily implemented in this method of SSI. In substructure approach, the soil–structure system is divided into two or more substructures. Each substructure is modelled independently and the general structure is formed by connecting these individual substructures through the interface of adjacent or other substructures. The substructure method is based on the principle of superposition.

The two most common soil models which are generally used for soil–structure interaction problems are winkler spring model and finite element models of an elastic continuum. In winkler spring method, soil medium is assumed to consist of a series of closely spaced springs on which the foundation slab lies. The springs are linear in nature and are dependent on the subgrade modulus (Wolf1985; Arya and Paul1977; Bowles1997). Elastic continuum model is a conceptual approach of physical representation of the infinite soil media (Rajasankar et al.2007; Tabatabaiefar and Massumi2010; Cakir2013). For the finite element model, the accuracy is valid to the extent of realistic estimate of the elastic modulus of the soil and Poisson's ratio (Chowdhury and

Dasgupta2009). Real progress in the area of three-dimensional soil–structure interaction has taken place with the advent of digital computers (Jayalekshmi et al.2011; Rajasankar et al.2007; Tabatabaiefar and Massumi2010; Cakir2013).

Only a few studies have been carried out on the SSI analysis of tall chimney structures under wind load compared to seismic load (Pour and Chowdhury2008; Jayalekshmi et al.2011). It is also found that limited research has been done in the area of SSI analysis of tall chimneys with piled raft foundation. In this parametric study, three-dimensional SSI analysis of reinforced concrete chimneys with piled raft foundation subjected to wind loads was carried out using finite element method based on the direct method of SSI. The equivalent static wind loads were computed as per IS: 4998(Part 1)-1992(2003). The linear elastic material behaviour of chimney, piled raft and the soil was assumed. Different responses in chimney such as lateral deflection, tangential moment, radial moment and base moment were evaluated incorporating SSI. These responses obtained from SSI analysis were compared with those obtained from the analysis of chimney with fixed base.

CHARACTERISTICS OF STRUCTURAL AND GEOTECHNICAL MODEL

Idealization of Chimney

Tall reinforced concrete chimneys of different elevations and base diameters were considered for the present study. Practical range of ratio of height to base diameter (slenderness ratio) of chimneys varies from 7 to 17 (Menon and Rao1997a). The chimney elevations of 100, 200, 300 and 400m were selected with slenderness ratios (H/D_b) of 7, 12 and 17. The taper ratio (ratio of top diameter to base diameter) and ratio of base diameter to thickness at bottom were considered as 0.6 and 35, respectively. The thickness at top of chimney was taken as 0.4 times the thickness at bottom but the minimum thickness at top was kept as 0.2m. All the above geometric parameters of chimney were selected based on the study conducted by Menon and Rao (1997a). Details of different geometric parameters of chimney are given in Table1. Linear elastic material behaviour of chimney was assumed in the study. M30 grade concrete and Fe 415 grade steel were selected as the materials for chimney. The modulus of elasticity for chimney was taken as 33.5Gpa as per IS: 4998(Part 1)-1992(2003). The Poisson's ratio and density of concrete were taken as 0.15 and 25kN/m^3, respectively, for chimney.

Table1: Geometric properties of chimney and piled raft

Chimney						Raft		Thickness(m)			Pile
Height H (m)	Slenderness ratio H/D_b	Diameter at base D_b (m)	Diameter at top D_t (m)	Thickness at base T_b (m)	Thickness at top $T_t=0.4T_b$ or 0.2m (m)	External diameter D_o (m)	Internal diameter D_i (m)	$D_o/t=12.5$	$D_o/t=17.5$	$D_o/t=22.5$	Total number of piles
100	7	14.5	8.7	0.5	0.2	30	9	2.4	1.71	1.33	79
	12	8.5	5.1	0.3	0.2	20	6	1.6	1.2	0.9	39
	17	6	3.6	0.2	0.2	14	4	1.12	0.8	0.62	18
200	7	29	17.4	0.82	0.35	60	16	4.8	3.42	2.7	275
	12	17	10.2	0.5	0.2	35	10	2.8	2	1.6	94
	17	12	7.2	0.35	0.2	26	6	2.08	1.5	1.2	45
300	7	43	25.8	1.3	0.5	96	20	7.68	5.5	4.3	784
	12	25	15	0.7	0.3	60	12	4.8	3.4	2.7	306
	17	18	10.8	0.5	0.2	42	8	3.36	2.4	1.9	151
400	7	57.5	34.5	1.7	0.7	140	20	11.2	8	6.3	1697
	12	33.5	20.1	1	0.4	86	16	6.88	5	3.9	641
	17	24	14.4	0.7	0.3	60	8	4.8	3.4	2.7	311

IDEALIZATION OF PILED RAFT FOUNDATION

Tall chimneys supported over piled raft foundation were considered. The raft of piled raft foundation was considered as annular with uniform thickness. The overall diameter of raft for a concrete chimney is typically 50% greater than the diameter of the chimney shaft at ground level (Turner2005). The ratio of outer diameter to thickness (D_o/t) of annular raft was taken as 12.5, 17.5 and 22.5 (Jayalekshmi et al.2011). RC friction piles of 20m length (l) and 1m diameter were considered. For friction piles, the optimum spacing recommended is 3dwheredis the diameter of the pile. Spacing (s) of 3densures that interference of stress zones of adjacent friction piles is minimum and results in a high group efficiency. Therefore,s/dof 3 was selected for the present study. Table1gives the details of different geometric parameters of raft and the total number of piles. Figure1shows the plan view of raft of piled raft foundation of 200m chimney (H/D_b=12). The linear elastic material behaviour was considered for piled raft foundation. The modulus of elasticity of 27.39Gpa was calculated corresponding to M30 grade concrete using the equation,Ec=5000fck----√Ec =5000fckas there is no IS code that provides the modulus values for piled raft foundation directly. Grade of steel was selected as Fe 415. The Poisson's ratio for piled raft foundation was taken as 0.15 and density of concrete were taken as 25kN/m^3.

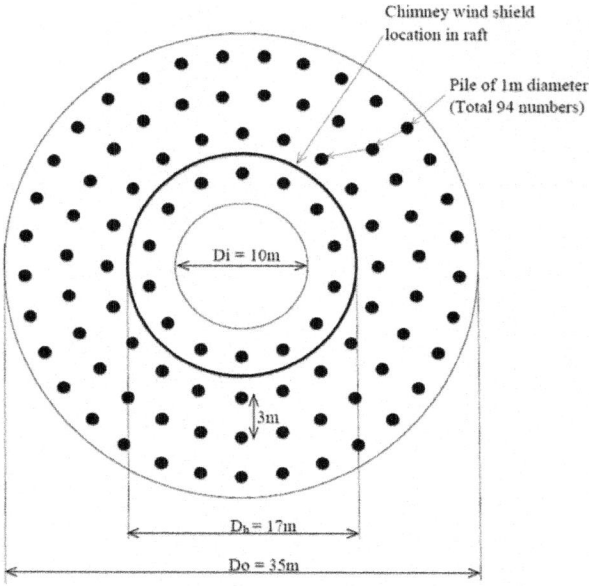

Figure1: Plan view of piled raft foundation of 200m chimney (H/D_b=12).

IDEALIZATION OF SOIL STRATUM

The soil is idealized by single homogeneous strata of 30m depth beneath the foundation. The bedrock was assumed to be at a depth of 30m for all chimneys. Wolf (1985) stated that the boundary of the soil should be placed at a sufficient distance from the structure where the static response has died out. Previous studies of SSI effect (Ghosh and Wilson1969; Rajasankar et al.2007; Tabatabaiefar and Massumi2010; Sáez et al.2011) considered the width of soil as 3–4 times the width of the foundation. In this study, the lateral boundary of soil was placed at a distance of four times the width of foundation. To study the effect of SSI, the properties of the soil stratum were varied. For this, four types of dry cohesionless soil, S1, S2, S3 and S4, were selected, which represent loose sand, medium sand, dense sand and rock, respectively. The properties of the soil stratum were defined by its mass density, elastic modulus and Poisson's ratio as per the references (Bowles1997; NEHRP1994). Coefficient of internal friction between the soil and the pile were taken as per Meyerhof from the foundation engineering book (Fang1991). The properties of the soil stratum are given in Table2.

Table2: Properties of the soil types

Soil type	Elastic modulus, $E(kN/m^2)$	Poisson's ratio,υ	Unit weight,$\gamma(kN/m^3)$	Angle of friction (°)
S1	108,000	0.4	16	30
S2	446,000	0.35	18	35
S3	1,910,000	0.3	20	40
S4	7,630,000	0.3	20	45

ESTIMATION OF ALONG-WIND AND ACROSS-WIND LOAD AS PER IS: 4998 (PART 1)-1992

There are two methods for estimating along-wind and across-wind loads for chimneys as per IS: 4998(Part 1)-1992(2003): simplified method and random response method. The chimneys are classified as class C structures located in terrain category 2 and subjected to a basic wind speed of 50m/s. According to IS: 875(Part 3)-1987(2003), terrain category 2 is an open terrain with well-scattered obstructions having heights generally between 1.5 and 10m.

Along-Wind Load

Simplified Method

The along-wind load or drag force per unit height (N/m) of the chimney at any level is calculated using the following equation:

$$F_z = p_z C_D d_z,$$

(1)

where p_z is the design wind pressure in N/m²at height z,zis the height of any section of chimney from top of foundation in m,C_D is the 0.8, drag co-efficient of chimney anddd_z is the diameter of chimney at heightzin m

Random Response Method

The along-wind load per unit height at any heightzon a chimney is calculated using the following equation:

$$F_z = F_{zm} + F_{zi},$$

(2)

where$F_{zm}$is the wind load in N/m height due to hourly mean wind (HMW) speed at heightzand$F_{zi}$is the wind load in N/m height due to the fluctuating component of wind at heightz.

$$F_{zm} = \bar{p}_z C_D d_z,$$

(3)

where $\bar{p}_z$ is the design pressure at heightz (N/m²) due to HMW speed, $\bar{P}_z = 0.6\bar{V}_z^2$, where $\bar{V}_z$ is the HMW speed in m/s.

$$F_{zi} = 3\frac{(G-1)}{H^2}\frac{z}{H}\int_0^H F_{zm}z\,dz,$$

(4)

where$G$is the gust factor which is calculated from the following equation:

$$G = 1 + g_f r\sqrt{B + \frac{SE}{\beta}},$$

(5)

where$g_f$is the peak factor defined as the ratio of the expected peak value to

RMS value of the fluctuating load,ris the twice the turbulence intensity,Bis the background factor indicating the slowly varying component of wind load fluctuation,Eis the measure of the available energy in the wind at the natural frequency of chimney,Sis the size reduction factor,βis the coefficient of damping of the structure andHis the total height of the chimney in m.

Across-Wind Load

Simplified Method

The amplitude of vortex excited oscillation perpendicular to direction of wind for any mode of oscillation shall be calculated by the following formula:

$$\eta_{oi} = \frac{\int_0^H d_z \phi_{zi} d_z}{\int_0^H \phi_{zi}^2 d_z} \times \frac{C_L}{4\pi S_n^2 K_{si}},$$

$$(6)$$

where$\eta_{oi}$is the peak tip deflection due to vortex shedding in theith mode of vibration in m,$C_L$is the 0.16, peak oscillatory lift coefficient,$K_{si}$is the mass damping parameter for theith mode of vibration, S_n=0.2, Strouhal number and$\phi_{zi}$is the mode shape function normalized with respect to the dynamic amplitude at top of the chimney in theith mode of vibration.

Periodic response of the chimney in theith mode of vibration is very strongly dependent on a dimensionless mass damping parameterK_{si}calculated by the following formula:

$$K_{si} = \frac{2m_{ei}\delta_s}{\sigma d^2},$$

$$(7)$$

where$m_{ei}$is the equivalent mass per unit length in kg/m in theith mode of vibration

$$m_{ei} = \frac{\int_0^H m_z \phi_{zi}^2 d_z}{\int_0^H \phi_{zi}^2 d_z},$$

$$(8)$$

where$\delta_s$is the logarithmic decrement of structural damping,σ=1.2kg/m³, mass density of air anddis the effective diameter taken as average diameter over the top 1/3rd height of the chimney in m.

The sectional shear force (F_{zoi}) and bending moment (M_{zoi}) at any heightzo, for theith mode of vibration, shall be calculated from the following equation:

$$F_{zoi} = 4\pi^2 f_i^2 \eta_{oi} \int_{zo}^{H} m_z \phi_{zi} dz$$

(9)

$$M_{zoi} = 4\pi^2 f_i^2 \eta_{oi} \int_{zo}^{H} m_z \phi_{zi}(z - zo) dz,$$

(10)

where f_i is the natural frequency of chimney in the ith mode of vibration in Hz and m_z is the mass per unit length of the chimney at section z in kg/m.

The fundamental mode of vibration is considered for computing the across-wind load. The fundamental natural frequencies of chimneys with fixed base are given in Table 3. In chimneys with fixed base, it is seen that as the height of chimney subsequently increases from 100 to 200 m, from 200 to 300 m and from 300 to 400 m, the natural frequency of chimney is reduced by 40–50, 25–35 and 22–26%, respectively.

Table3: Natural frequency of chimney with fixed base and flexible base

H(m)	H/D_b	Natural frequency of chimney with fixed base (Hz)	Natural frequency of chimney due to the effect of SSI (Hz)											
			D_o/t=12.5				D_o/t=17.5				D_o/t=22.5			
			S1	S2	S3	S4	S1	S2	S3	S4	S1	S2	S3	S4
100	7	1.339	1.018	1.169	1.259	1.313	0.950	1.127	1.241	1.306	0.907	1.102	1.229	1.301
	12	0.688	0.584	0.628	0.655	0.673	0.557	0.612	0.649	0.671	0.530	0.598	0.645	0.669
	17	0.427	0.389	0.407	0.418	0.425	0.375	0.400	0.415	0.424	0.365	0.395	0.414	0.423
200	7	0.670	0.547	0.604	0.639	0.659	0.511	0.584	0.631	0.656	0.487	0.573	0.626	0.654
	12	0.400	0.346	0.374	0.390	0.399	0.330	0.366	0.387	0.398	0.319	0.361	0.385	0.397
	17	0.258	0.226	0.240	0.247	0.251	0.219	0.235	0.245	0.250	0.213	0.233	0.244	0.250
300	7	0.454	0.392	0.420	0.439	0.451	0.366	0.406	0.433	0.449	0.347	0.398	0.430	0.447
	12	0.261	0.240	0.250	0.257	0.261	0.228	0.245	0.255	0.261	0.220	0.241	0.254	0.260
	17	0.193	0.181	0.188	0.192	0.194	0.174	0.185	0.190	0.194	0.170	0.183	0.190	0.193
400	7	0.336	0.303	0.318	0.322	0.335	0.285	0.309	0.325	0.334	0.272	0.303	0.322	0.333
	12	0.201	0.187	0.194	0.198	0.201	0.179	0.190	0.197	0.201	0.172	0.187	0.195	0.200
	17	0.142	0.135	0.139	0.141	0.143	0.130	0.136	0.141	0.143	0.126	0.135	0.140	0.143

Random Response Method

Calculation of across-wind load is made by first calculating the peak response amplitude at the specified mode of vibration (usually the first or second). The taper of chimneys with slenderness ratio (H/D_b) that equals 7 was more than 1 in 50 and that of other chimneys was less than 1 in 50. Taper is defined as $\{2\ (d_{av} - d_{top})/H\}$ where d_{av} is the average outer diameter over the top half of chimney and d_{top} is the outer diameter at the top of chimney.

For chimney with little or no taper (average taper over the top one-third height is less than or equal to 1 in 50), the modal response, at a critical wind speed is calculated by the following formula:

$$\eta_{oi} = \frac{\dfrac{1.25\bar{C}_L d\phi_{zi} H_i}{\pi^2 S_n^2} \times \sigma d^2 \sqrt{\dfrac{\sqrt{\pi L}}{2(\cap+2)}}}{m_{ei}\left(\dfrac{1}{H}\int_0^H \phi_{zi}^2 dz\right)^{\frac{1}{2}}\left(\beta - \left(\dfrac{k_a \sigma d^2}{m_{ei}}\right)\right)^{\frac{1}{2}}},$$

(11)

where $\cap$ is the equivalent aspect ratio=H/d, H is the height of chimney in m, d is the average diameter over the top 1/3rd height of chimney in m, $\bar{C}_L$ =0.12, RMS lift coefficient, L=1, correlation length in diameters k_a=0.5, aerodynamic damping co-efficient.

For the chimney which is significantly tapered (average taper over the top one-third height is more than 1 in 50), the modal response is calculated by the following formula:

$$\eta_{oi} = \frac{\sigma \bar{C}_L d_{ze}^4 \phi_{zei} \phi_{zi} H_i \left(\frac{\pi L}{2t}\right)^{\frac{1}{2}}}{2\pi^2 S_n^2 m_{ei} \int_0^H \phi_{zi}^2 dz \left(\beta - \frac{k_a \sigma d^2}{m_{ei}}\right)^{\frac{1}{2}}},$$

(12)

where, z_{ei} is the height in meter at which $d_z^4 \phi_{zi}/\sqrt{t}$ is maximum in the ith mode of vibration in m

$$t = \left\{\frac{-\delta}{\delta_z} d_z + \frac{\alpha d_z}{z}\right\}_{z=zei}$$

(13)

where, α is the power law exponent. For terrain category 2, the value of α is 0.14.

The across-wind and along-wind loads for tall RC chimneys were obtained by both simplified and random response methods. The base moments of chimney due to across-wind and along-wind loads are shown in Table4. It is found that the base moment of the chimney computed from simplified method is lower than that from the random response method when it is subjected to along-wind load. The variation of base moment obtained from random response method and simplified method decreases with increase in chimney elevation. In the lower elevation chimney (H=100m) with differentH/D_bratios, the variation of base moment of 110–152% is found between the two methodologies, whereas in higher elevation chimneys (H=400m) this variation between the two methods is 52–92%. In the case of across-wind load, the higher base moment of chimney is obtained from simplified method. The variation of base moment of all chimneys estimated from the two methods ranges in 18–56%. The above variation is caused by the difference in the value ofη_{oi}.

Table4: Base moment of chimney due to along and across wind load as per IS: 4998 (Part 1)-1992

Height of chimney,H(m)	Slenderness ratio (H/D_b)	Base moments (kNm)			
		Along wind		Across wind	
		Simplified method	Random response method	Simplified method	Random response method
100	7	59,449	124,795	731,018	503,019
	12	34,849	79,420	38,787	20,556
	17	24,600	62,045	5316	2335
200	7	540,322	996,474	5,498,938	4,535,521
	12	316,740	629,786	393,407	259,215
	17	223,581	485,981	57,683	30,173
300	7	1,957,856	3,335,420	17,553,686	14,316,742
	12	1,138,288	2,124,915	1,132,171	739,151
	17	819,568	1,638,583	230,940	132,810
400	7	4,930,240	7,481,986	39,879,581	32,339,199
	12	2,872,401	5,087,830	7,841,886	5,082,837
	17	2,057,839	3,958,270	515,128	285,123

The wind load for which the maximum base moment of the chimney is obtained is selected out of the across-wind and along-wind loads for SSI analysis. This wind load is applied to the finite element model of chimney at various locations along the height of chimney. It is also found that for stocky chimneys (H/D_b=7), the maximum base moment is obtained due to across-wind load and for slender chimneys (H/D_b=17), the base moment of chimney is maximum due to along-wind load. It is also observed that the maximum base moment of 100, 200 and 300m chimneys withH/D_b=12 is obtained when it is subjected to along-wind load, whereas across-wind load causes maximum base moment in 400m tall chimney.

FINITE ELEMENT MODEL OF CHIMNEY–PILED RAFT–SOIL SYSTEM

The three-dimensional finite element analysis of chimney with fixed base and integrated chimney–piled raft–soil system was carried out using the finite element software ANSYS. The chimney and the raft were modelled using four-noded SHELL63 element, which has both bending and membrane capabilities. This element has six degrees of freedom at each node. Eight-noded SOILD45 elements with three translational degrees of freedom at each node were used for the three-dimensional modelling of the soil and the pile. The surface–surface contact elements were used to model the interaction between pile and soil. The pile surface was established as "target" surface (TARGE170), and the soil surface contacting the pile as "contact" surface (CONTAC174); these two surfaces constitute the contact pair. The coefficient of friction was defined between contact and target surfaces and is given in Table2. The lateral movements at the soil boundaries were restrained. All movements were restrained at bed rock level. The nodes at the interface of bottom of raft and top of soil were completely coupled.

The chimney shell was discretised with element of 2m size along height and with divisions of 7.5° in the circumferential direction. The diameter and thickness of chimney were varied linearly along the entire height. The pile was discretised with 14 elements of same size along the length of pile.

Elastic continuum approach was adopted for modelling the soil. The material properties such as elastic modulus, Poisson's ratio and density for the three-dimensional soil stratum are given in Table2. The integrated chimney–foundation–soil system was analysed based on direct method of SSI by assuming the linear elastic behaviour of the whole system.

The wind load computed as per IS: 4998(Part 1)-1992(2003) was applied in the chimney as point loads at 10m intervals along its height after suitably averaging the load above and below each section. The gravity load was also applied to the SSI model. Finite element model of 100m chimney (H/D_b=7) with fixed base subjected to across-wind load is shown in Fig.2. Three-dimensional finite element model of the integrated chimney–piled raft–soil system was generated using the ANSYS software and is shown in Fig.3. The finite element model of piled raft foundation and that of a single pile are shown in Fig.4.

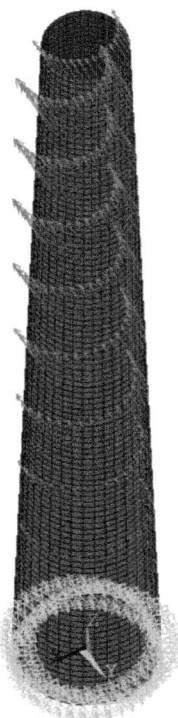

Figure2: Finite element model of 100m chimney (H/D$_b$=7) with fixed base.

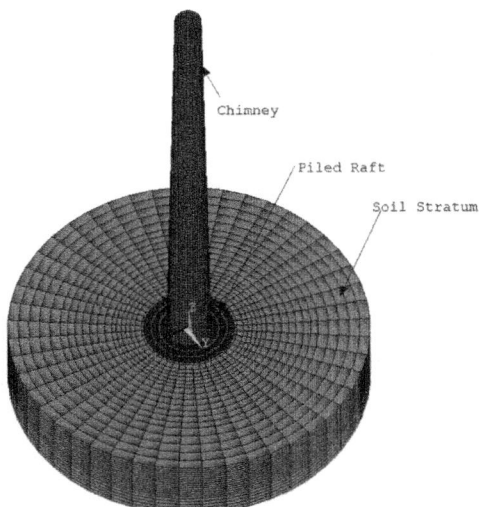

Figure3: Finite element model of integrated 200m chimney (H/D$_b$=12)–piled raft–soil system.

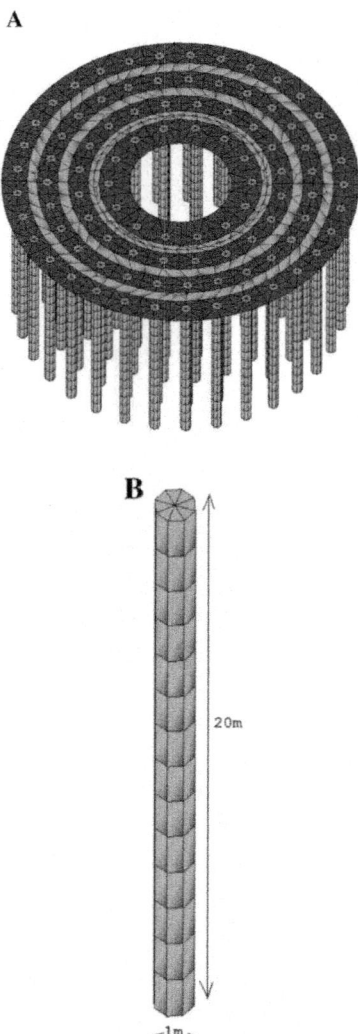

Figure4: Finite element model ofapiled raft andbpile.

The responses of chimney in terms of lateral deflection, base moment, tangential moments and radial moments were investigated. The responses of chimney obtained from the SSI analysis of chimney–piled raft system were compared with that obtained from chimney with fixed base. The results obtained from finite element analysis of chimney with fixed base are designated as "Fixed" in graphs and tables. The percentage variation of maximum values of the moments in the chimney considering SSI from those obtained from the finite element analysis of chimney with fixed base was computed. The effect of SSI was studied by considering different parameters such as flexibility of soil, stiffness of the raft of piled raft foundation, slenderness ratio of the chimney and chimney elevation.

RESULTS AND DISCUSSIONS

Finite element analysis was conducted on 144 integrated three-dimensional chimney–piled raft–soil systems under wind loads applied along the height of chimney to study the effect of SSI in wide range of chimneys with piled raft foundations subjected to wind loads. The responses of chimney such as lateral deflection, base moment, tangential and radial moments, etc. were analysed. The variation of response of chimney with flexible base from that of chimney with fixed base was computed. The maximum response is obtained at the leeward side of the chimney and, therefore, the different responses at the leeward side of chimney are shown in the following graphs. The effect of flexibility of soil, stiffness of raft of piled raft foundation, slenderness ratio of chimney and of chimney elevation on the above variation is studied.

EFFECT OF FLEXIBILITY OF SOIL

To study the effect of SSI, four types of soils were selected namely S1, S2, S3 and S4 representing loose sand, medium sand, dense sand and rock respectively. The natural frequency, lateral deflection and moments of the chimney were evaluated considering fixed base and flexible base for the chimney.

Variation of Natural Frequency of Chimney

The natural frequencies obtained from SSI analysis were compared with that obtained from the analysis of chimney with fixed base and is given in Table3. The fundamental natural frequency obtained from chimney with fixed base is higher than that obtained from SSI analysis. The variation of fundamental frequency of chimney with flexible base from that of chimney with rigid base is more for 100m chimney (H/D_b=7 andD_o/t=22.5) resting on soil type S1 and the variation is 32.26%. The variations of same chimney resting on soil types S2, S3 and S4 are 17.69, 8.21 and 2.83%, respectively. It is seen that the percentage variation of natural frequency between that of flexible base and fixed base conditions of chimney is considerable for soil type S1 and S2. Due to interaction with stiff soil types S3 and S4, the above variation is less than 10%. Therefore, it can be inferred that the SSI effect is prominent on flexible soil types rather than stiff soil types.

Variation of Lateral Deflection of Chimney

The lateral deflection of chimney is obtained from the analysis of chimney with piled raft foundation considering the SSI effect and fixity at the base of chimney. The contour of lateral displacement of 200m chimney (H/D_b=7 and D_o/t=22.5) resting on the four soil types and that of same chimney with fixed base are shown in Fig.5. The lateral deflection along the height of 100, 200, 300 and 400m chimney (H/D_b=12) with piled raft (D_o/t=22.5) foundation obtained from fixed base analysis and SSI analysis is shown in Fig.6. It is found that the deflection of chimney increases with increase in flexibility of soil. The deflection is maximum at the tip of chimney for all cases. The tip deflection is tabulated in Table5. It is seen that in general, the tip deflection of chimney obtained from the analysis of chimney with fixed base is lower than that obtained from the SSI analysis. Maximum increase in tip deflection of 89% is found for 100m chimney (H/D_b=7) with piled raft (D_o/t=22.5) under flexible soil type S1 from the chimney with fixed base. For the same chimney–foundation system, the maximum variation of tip deflection of chimney resting on soil type S2, S3 and S4 from that of chimney with fixed base is 39, 17 and 6%, respectively. The soil–structure interaction studies are relevant for chimneys resting on soil types S1 and S2 as the variation of tip deflection from fixed base analysis is significant for all chimneys considered.

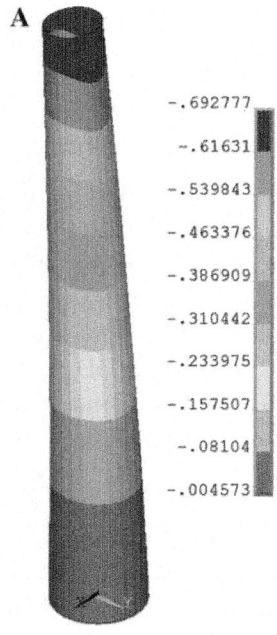

A

-.692777
-.61631
-.539843
-.463376
-.386909
-.310442
-.233975
-.157507
-.08104
-.004573

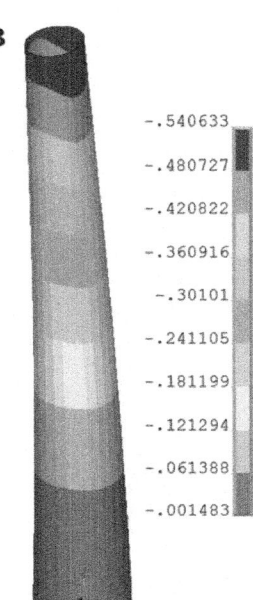

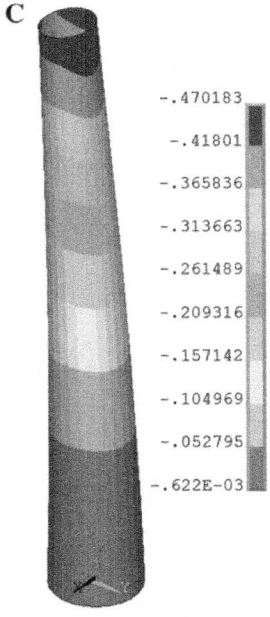

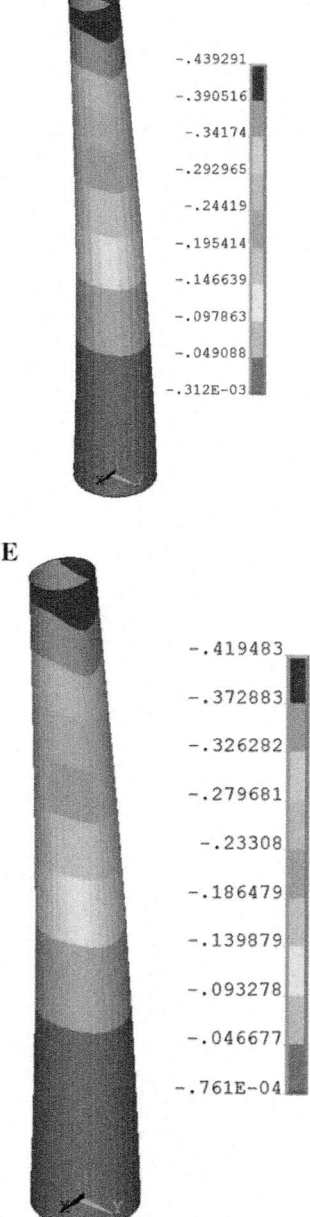

Figure5: Contour of lateral deflection (m) of 200m chimney (H/D$_b$=7 andD$_o$/ t=22.5) resting on soil typesaS1,bS2,cS3,dS4 andefixed base.

A

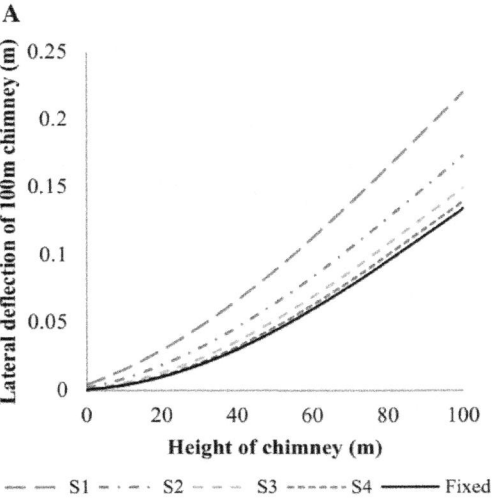

B

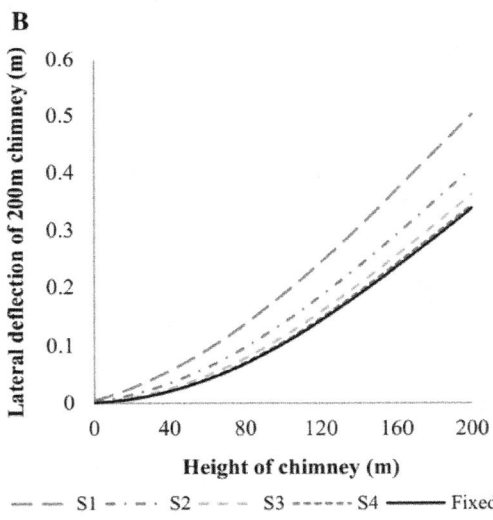

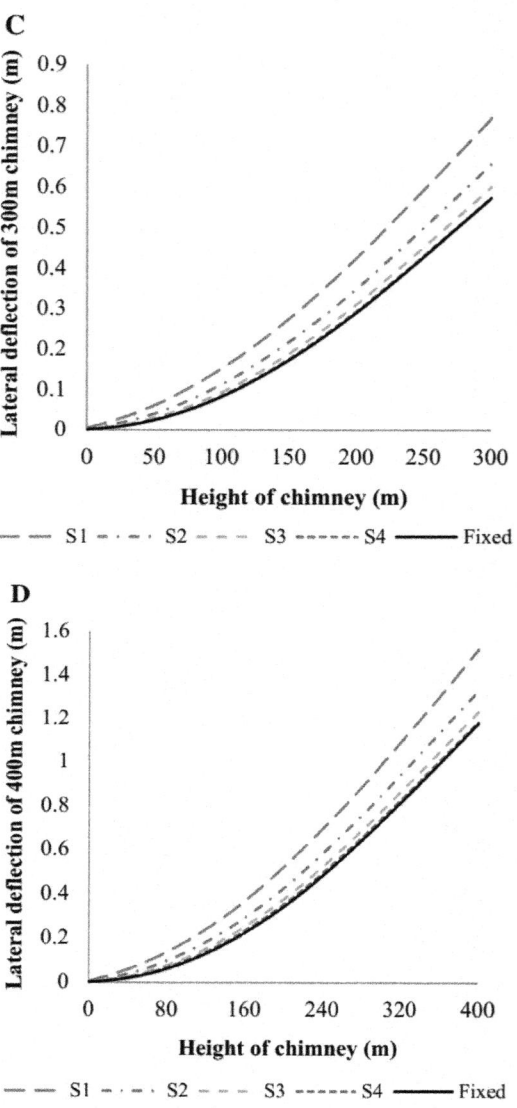

Figure6: Lateral deflection ofa100m,b200m,c300m,d400m chimney (H/D$_b$=12 andD$_o$/t=22.5) with flexible and fixed base.

Table5: Lateral deflection of chimney with fixed base and flexible base

H(m)	H/D_b	Tip deflection of chimney with fixed base (m)	Tip deflection of chimney due to chimney-piled raft–soil interaction (m)											
			$D_o/t=12.5$				$D_o/t=17.5$				$D_o/t=22.5$			
			S1	S2	S3	S4	S1	S2	S3	S4	S1	S2	S3	S4
100	7	0.18	0.28	0.23	0.20	0.19	0.31	0.24	0.21	0.19	0.34	0.25	0.21	0.19
	12	0.13	0.18	0.16	0.15	0.14	0.20	0.17	0.15	0.14	0.22	0.17	0.15	0.14
	17	0.41	0.52	0.47	0.44	0.43	0.56	0.49	0.47	0.43	0.60	0.50	0.45	0.43
200	7	0.42	0.58	0.50	0.46	0.44	0.64	0.53	0.47	0.44	0.70	0.55	0.48	0.44
	12	0.34	0.44	0.39	0.36	0.35	0.48	0.40	0.36	0.35	0.51	0.41	0.37	0.35
	17	1.00	1.08	0.97	0.92	0.89	1.16	1.01	0.93	0.89	1.22	1.03	0.94	0.90
300	7	0.59	0.74	0.67	0.63	0.61	0.84	0.71	0.64	0.61	0.90	0.73	0.65	0.61
	12	0.57	0.66	0.61	0.59	0.57	0.72	0.64	0.60	0.57	0.77	0.66	0.60	0.57
	17	1.66	1.89	1.77	1.71	1.67	2.01	1.83	1.73	1.68	2.11	1.87	1.74	1.68
400	7	0.75	0.79	0.73	0.69	0.66	0.87	0.76	0.70	0.67	0.95	0.79	0.71	0.67
	12	1.18	1.34	1.27	1.23	1.20	1.44	1.32	1.24	1.20	1.54	1.35	1.25	1.21
	17	2.09	2.36	2.25	2.18	2.14	2.52	2.32	2.20	2.14	2.66	2.37	2.22	2.15

Variation of Tangential Moment in Chimney

The tangential bending moments in chimney is evaluated from the SSI analysis and fixed base analysis of chimney. The contours of tangential moment of 200m chimney (H/D_b=7 andD_o/t=22.5) supported on different soil types and with fixed base are shown in Fig.7. The tangential moment at various locations along the height of 100m chimney in the leeward side is shown in Fig.8for the SSI and fixed base cases. It is observed that the maximum tangential moment in chimney with fixed base is obtained at a height of H/3 from the top. The wind load intensity is more in this region of chimney. It is seen that tangential moments are high at the bottom of the chimney also but not the maximum for the case of chimney with fixed base. The maximum tangential moment is obtained atH/3m from the top of chimney in the case of chimney founded on supporting soil type S4 also which is the same as in the above said case of chimney with fixed base. It is observed that the maximum tangential moment of chimney is obtained at the base of chimney when it rests on soil type S1 and S2. It is found that the higher moments occur at the base as well as atH/3m from the top of chimney when it is supported on soil type S3. The above observations correspond to stocky chimneys (H/D_b=7). All other chimneys show the maximum tangential moment at the base of chimney itself due to the SSI effect.

A

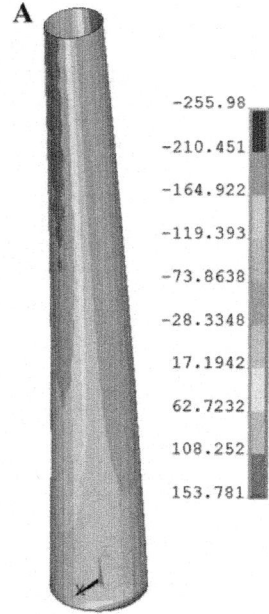

$$-255.98$$
$$-210.451$$
$$-164.922$$
$$-119.393$$
$$-73.8638$$
$$-28.3348$$
$$17.1942$$
$$62.7232$$
$$108.252$$
$$153.781$$

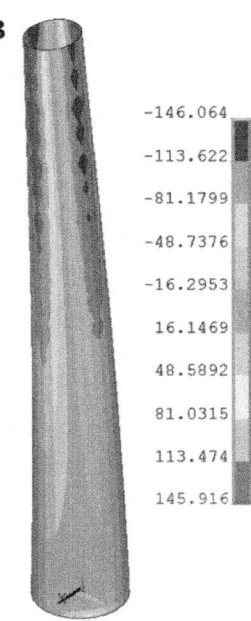

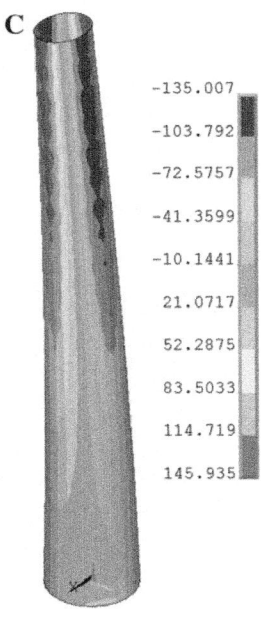

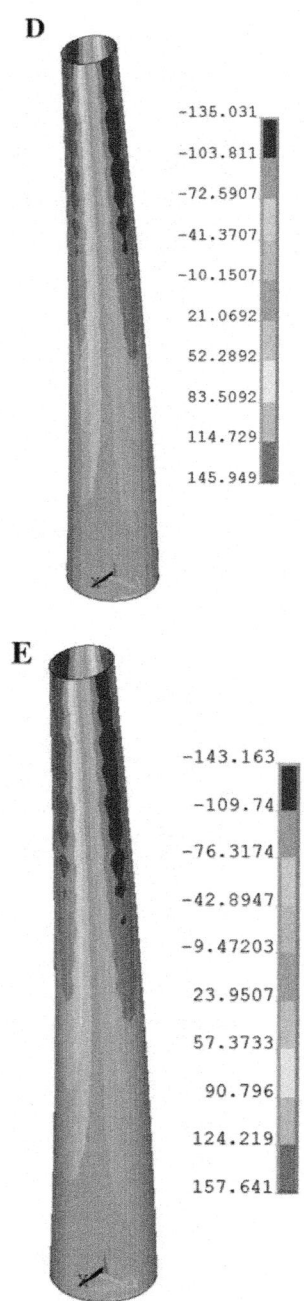

Figure7: Contour of tangential moment (kNm) of 200m chimney (H/D$_b$=7 andD$_o$/t=22.5) resting on soil typesaS1,bS2,cS3,dS4 andefixed base.

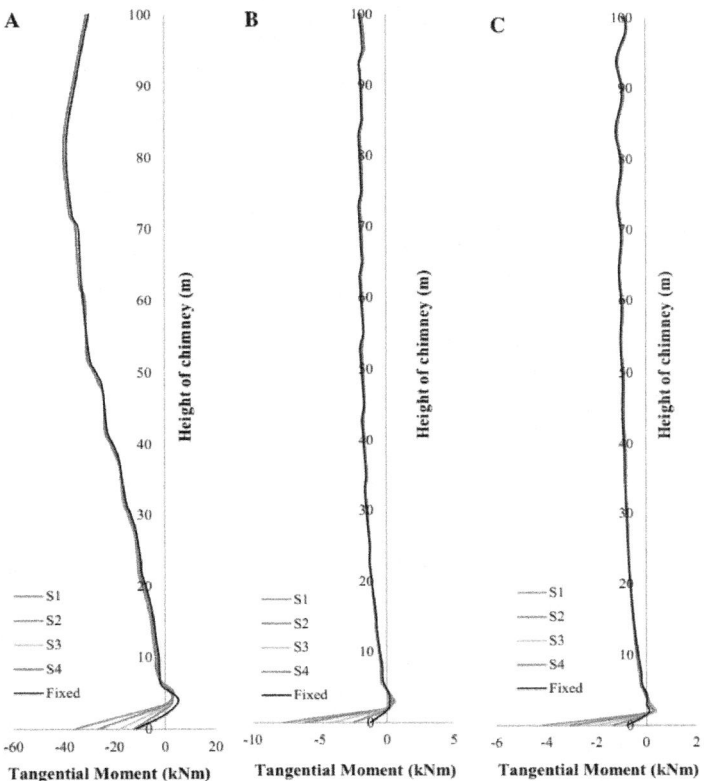

Figure8: Tangential moment of 100m chimney ($D_o/t=12.5$) with differentH/D_bratios ofa7,b12 andc17.

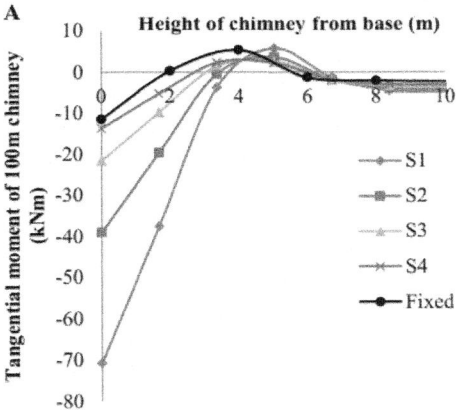

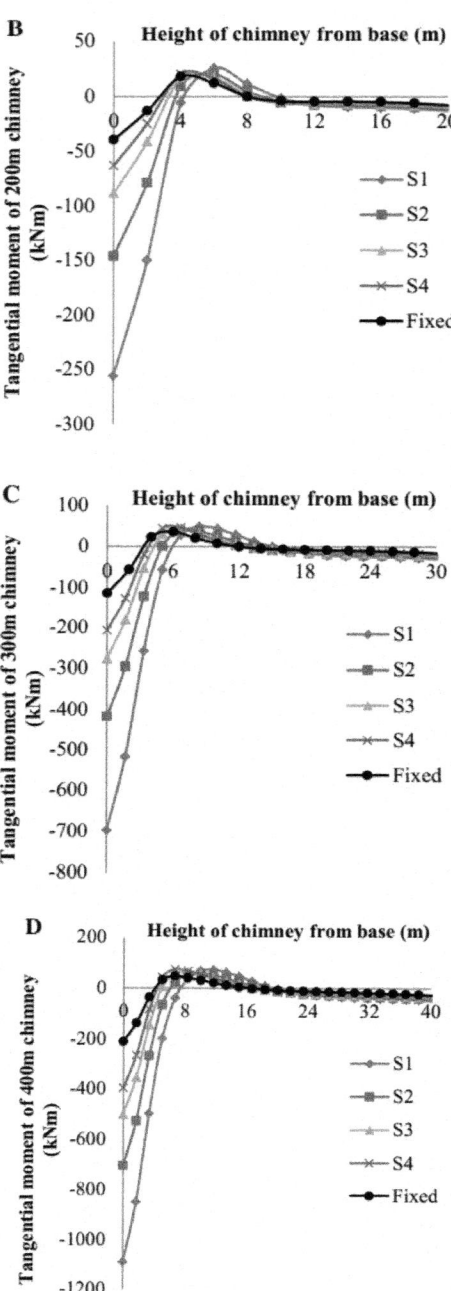

Figure9: Tangential moment ofa100m,b200m,c300m,d400m chimney (H/D$_b$=7 andD$_o$/t=22.5) with flexible and fixed base.

Table6: Variation of tangential moment of chimney due to chimney–piled raft–soil interaction

H(m)	H/D_b	Tangential moment of chimney with fixed base (kNm)	Percentage variation of tangential moment of chimney due to chimney–piled raft–soil interaction (%)											
			$D_o/t=12.5$				$D_o/t=17.5$				$D_o/t=22.5$			
			S1	S2	S3	S4	S1	S2	S3	S4	S1	S2	S3	S4
100	7	38.845	2.66	2.67	2.69	2.71	45.57	2.65	2.68	2.70	82.05	2.63	2.68	2.70
	12	2.4374	217.85	153.10	72.23	4.75	425.31	262.28	102.46	9.01	699.21	377.35	123.66	13.50
	17	1.8456	125.55	65.24	11.14	-3.23	268.41	116.68	16.91	-3.23	378.22	132.10	7.92	-3.23
200	7	139.16	0.88	-2.98	-2.97	-2.96	49.03	-3.00	-2.98	-2.97	83.92	4.94	-2.98	-2.97
	12	9.0026	216.70	121.20	43.14	2.08	376.90	182.63	58.85	2.08	482.00	216.33	68.24	5.17
	17	4.7856	157.46	103.02	46.78	30.51	332.86	184.39	73.89	30.51	471.95	238.66	91.99	30.51
300	7	276.91	51.34	14.66	7.72	7.73	107.66	36.46	7.71	7.73	151.89	51.11	7.70	7.72
	12	17.511	284.49	179.68	100.71	51.97	452.18	248.81	122.72	58.54	578.03	292.92	135.48	62.78
	17	9.5946	293.17	199.48	121.11	66.67	501.42	293.01	153.62	76.15	671.80	358.28	173.53	81.52
400	7	470.81	149.70	116.52	107.38	107.39	194.80	135.59	107.37	107.38	231.58	149.67	107.36	107.38
	12	56.109	275.47	180.63	111.52	67.11	420.18	244.15	134.08	74.01	554.28	294.59	149.64	79.24
	17	22.272	325.77	228.15	150.76	97.07	537.30	327.38	189.42	109.16	713.13	399.78	213.59	116.15

From the SSI analysis, it is found that the tangential bending moment in chimney increases with increase in flexibility of soil. It is also noticed that the effect of soil flexibility on tangential moment is negligible beyondH/20m height from the base of chimney. The variation of tangential moment of chimney with flexible base from that with fixed base is seen only up to the height ofH/20 from the base of the chimney. These variations for a height ofH/10m from the base of 100, 200, 300 and 400m chimneys (H/D_b=7 andD_o/ t=22.5) are shown in Fig.9. The percentage variation of maximum tangential moment of chimney with flexible base from that with fixed base is tabulated in Table6. The maximum variation of 713% of tangential moments is observed for very slender 400m chimney with flexible raft (H/D_b=17 andD_o/t=22.5) founded on loose sand and the corresponding variations when it interacts with S2, S3 and S4 soil types are 400, 214 and 116%, respectively.

Variation of Radial Moment in Chimney

The contour of the radial moment in chimney (H=200m,H/D_b=7 andD_o/ t=22.5) resting on four different soil types and that with fixed base are shown in Fig.10. It is seen that the maximum radial moment is obtained at the base of the chimney for all the analysis cases considered with and without SSI effect. The radial moments up to a height ofH/10 from the base of chimneys (H/D_b=7 andD_o/t=22.5) are shown in Fig.11. The radial bending moment in chimney increases with increase in flexibility of soil. It is also observed that the variation of radial moments of chimney due to different supporting soil conditions is seen only up to a height ofH/40 from the base of the chimney. The effect of soil flexibility on radial moments is negligible beyond thisH/40 height from the base of chimney. It is inferred that the state of stress developed at the base of chimney modelled with piled raft foundation and surrounding soil is different from that in the case of fixity at base due to the interaction among these three components. This effect naturally decays after a particular height above the base as the height of chimney is very large in comparison with the diameter at base. Hence the response variation at base for a distance of at least 1/40th of the height of chimney should be considered for safe design.

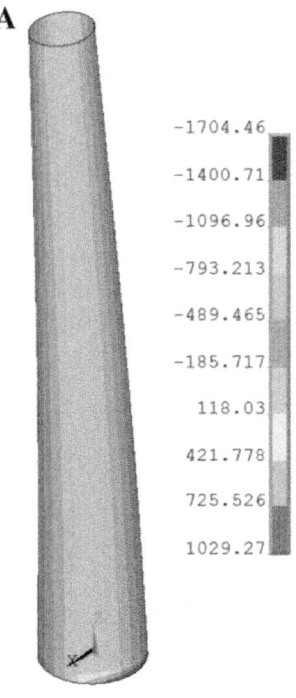

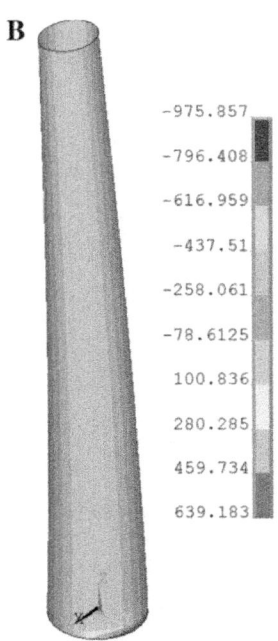

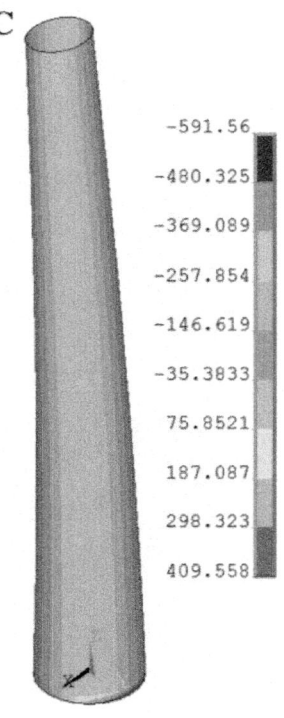

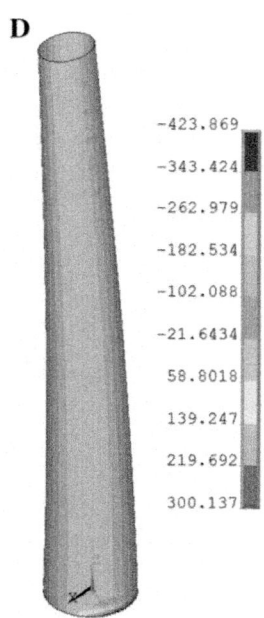

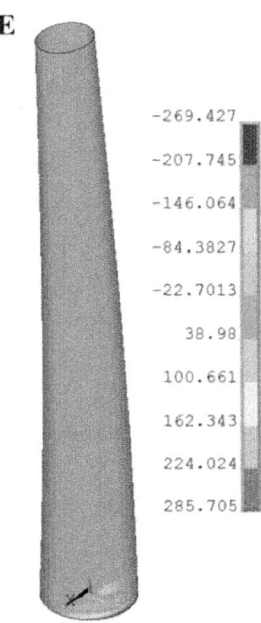

Figure10: Contour of radial moment (kNm) of 200m chimney (H/D$_b$=7 andD$_o$/ t=22.5) resting on soil typesaS1,bS2,cS3,dS4 andefixed base.

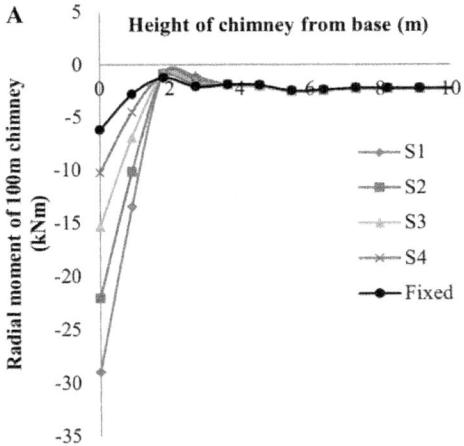

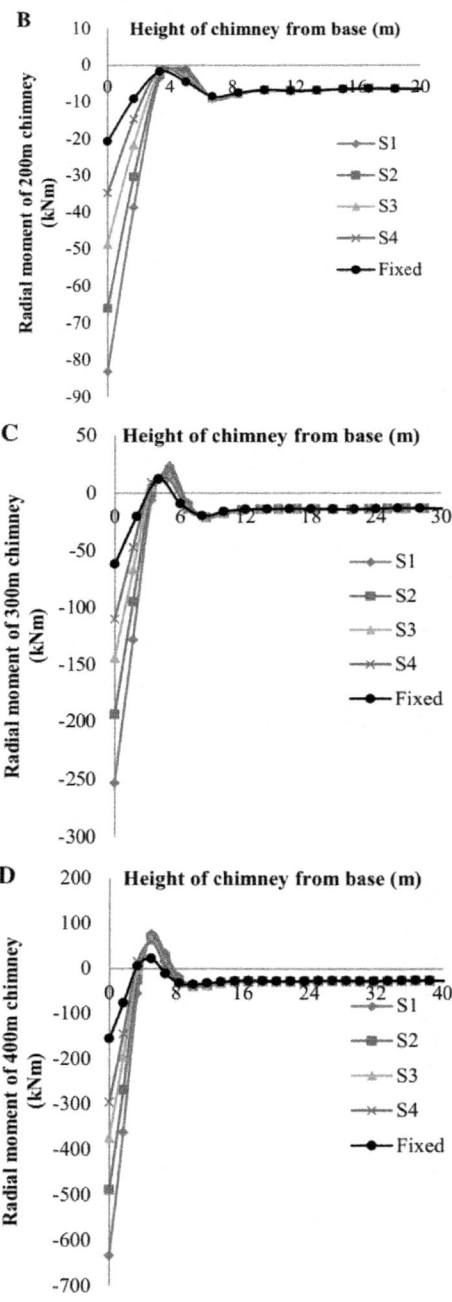

Figure11: Radial moment of a100m, b200m, c300m, d400m chimney (H/D$_b$=17) with flexible and rigid base.

Table7: Variation of radial moment of chimney due to chimney–piled raft–soil interaction

H(m)	H/D_b	Radial moment of chimney with fixed base (kNm)	Percentage variation of radial moment of chimney due to chimney–piled raft–soil interaction (%)											
			$D_o/t=12.5$				$D_o/t=17.5$				$D_o/t=22.5$			
			S1	S2	S3	S4	S1	S2	S3	S4	S1	S2	S3	S4
100	7	78.37	206.30	128.19	57.29	10.43	372.91	193.33	75.19	15.86	490.47	230.62	86.59	20.25
	12	8.75	488.95	370.65	222.86	99.58	867.05	569.79	277.63	106.67	1367.97	780.89	315.82	113.34
	17	6.20	367.28	255.41	147.16	65.14	661.40	382.00	177.69	69.85	901.82	451.73	185.98	72.43
200	7	269.43	248.96	166.92	95.33	50.15	413.34	228.56	111.05	54.56	532.56	262.13	119.47	57.24
	12	29.70	536.84	346.06	190.66	97.93	851.11	464.89	219.52	106.52	1055.53	528.73	235.67	113.13
	17	20.80	299.69	217.96	134.37	67.78	559.58	337.88	173.60	80.47	765.29	417.43	199.38	89.14
300	7	779.29	257.57	171.17	107.48	68.54	389.10	221.84	124.27	74.18	492.31	255.91	135.70	77.87
	12	85.71	423.30	281.37	174.74	109.29	647.37	373.15	203.60	117.81	815.01	431.55	220.27	123.41
	17	61.85	308.83	213.31	133.69	78.62	519.09	307.83	166.73	88.45	691.55	374.50	187.54	94.35
400	7	1427.90	228.87	156.15	106.78	76.59	327.07	197.53	122.56	81.90	407.16	228.10	133.06	85.08
	12	320.69	337.84	227.73	147.66	96.39	504.07	300.39	173.25	104.06	657.80	358.01	190.79	109.85
	17	156.34	304.84	212.87	140.22	90.10	501.14	304.42	175.52	100.91	663.91	371.08	197.44	107.07

The maximum radial bending moments in chimney with and without SSI effect are tabulated in Table7. Unlike tangential moments in chimney, the maximum radial moment of lower elevation chimneys (H=100m andH/D_b=12) with piled raft foundation having thin raft resting on soil type S1 is increased by 11–14 times of that of chimney with fixed base. This is the highest variation of radial moment of chimney with flexible base from that of chimney with fixed base obtained from all the analysis under consideration. Similarly, for this chimney supported on soil types S2, S3 and S4, the maximum variation is eight, four and two times of that of chimney with fixed base. The effect of SSI on the radial moment in chimney is more for lower elevation chimneys with a flexible raft while interacting with loose sand.

Variation in the Base Moment of Chimney

The base moment of chimney was computed for along-wind load and across-wind load according to IS: 4998(Part 1)-1992(2003) based on two methods. The wind load which caused the maximum base moment was applied to the SSI system and for this lateral load the base moment of chimney with flexible base was evaluated. The base moment obtained from the chimney with flexible base is compared with that obtained from the chimney with fixed base. The variations of base moment evaluated from both cases are shown in Table8. It is found that the base moment of chimney increases with increase in stiffness of supporting soil. It is also seen that the base moment is maximum in chimneys with fixed base as compared to flexible base. The base moment of the 100m chimney evaluated from the SSI analysis of chimney (H/D_b=17,D_o/t=22.5 and soil type S4) is decreased by 64% from that estimated from the chimney with fixed base. This is the minimum variation found between chimneys of flexible and fixed base. The maximum variation among both the cases is seen for a 400-m chimney (H/D_b=7,D_o/t=22.5 and soil type S4) and the reduction is 97%. It is seen that the base moment of chimney obtained from the chimney–piled raft system resting on rock is much less than that obtained from the chimney with fixed base.

Table8: Variation of base moment of chimney due to chimney-piled raft–soil interaction

H(m)	H/D$_b$	Base moment of chimney with fixed base (kNm)	Percentage variation of base moment of chimney due to chimney-piled raft–soil interaction (%)											
			D$_0$/t=12.5				D$_0$/t=17.5				D$_0$/t=22.5			
			S1	S2	S3	S4	S1	S2	S3	S4	S1	S2	S3	S4
100	7	731,018	−99.39	−98.20	−95.45	−90.96	−98.92	−97.16	−93.39	−86.90	−98.52	−96.23	−91.25	−82.79
	12	79,420	−99.05	−97.30	−93.43	−87.80	−98.43	−95.97	−91.07	−83.63	−97.54	−94.30	−88.12	−77.56
	17	62,045	−87.78	−85.84	−83.79	−80.41	−78.47	−78.30	−77.18	−72.52	−72.23	−72.10	−70.92	−64.53
200	7	5,498,938	−99.88	−99.45	−98.41	−96.45	−99.79	−99.10	−97.54	−94.48	−99.69	−98.75	−96.62	−92.38
	12	629,786	−99.58	−98.63	−96.52	−93.02	−99.25	−97.83	−94.96	−89.95	−98.99	−97.20	−93.53	−87.20
	17	485,981	−99.28	−97.86	−94.40	−88.63	−98.69	−96.48	−91.64	−83.48	−98.20	−95.36	−89.17	−78.88
300	7	17,553,686	−99.93	−99.61	−98.78	−97.32	−99.88	−99.37	−98.20	−96.14	−99.81	−99.14	−97.60	−94.94
	12	2,124,915	−99.84	−99.30	−97.97	−95.62	−99.69	−98.80	−96.92	−93.59	−99.54	−98.38	−95.98	−91.77
	17	1,638,583	−99.68	−98.86	−96.83	−93.33	−99.38	−98.06	−95.21	−90.33	−99.11	−97.36	−93.72	−87.51
400	7	39,879,581	−99.97	−99.80	−99.31	−98.44	−99.95	−99.68	−98.98	−97.76	−99.93	−99.56	−98.66	−97.07
	12	7,841,886	−99.91	−99.57	−98.66	−97.01	−99.84	−99.28	−98.00	−95.72	−99.74	−98.99	−97.31	−94.37
	17	3,958,270	−99.84	−99.31	−97.96	−95.53	−99.68	−98.78	−96.82	−93.44	−99.51	−98.32	−95.83	−91.55

EFFECT OF STIFFNESS OF RAFT

The effect of stiffness of raft of piled raft foundation was investigated by considering three different D_o/t ratios (D_o/t=12.5, 17.5, and 22.5) for the raft. It is seen that the response in chimney such as lateral deflection, tangential and radial moments and base moments increase with increase in D_o/t ratio. The stiffness of the foundation is less for higher D_o/t ratios of the raft and, therefore, the bending of chimney will be more at the base when it is subjected to the lateral wind load. The maximum variation in lateral deflection of chimney with flexible base from that of fixed base is observed for the 100-m chimney (H/D_b=7) resting on loose sand and the variations for D_o/t ratios of 12.5, 17.5 and 22.5 are 56, 72 and 89%, respectively. The representative figures of tangential and radial bending moments in the 100-m chimney for different Do/t ratios are shown in Fig.12. It is also found that the variation of moments in chimney with respect to different D_o/t ratios is only seen for a few metre heights (H/10m) from the base of the chimney. There is no effect of stiffness of raft on the moment response in chimney beyond this height. The tangential moment of chimney is increased by four times of that of fixed base due to SSI for D_o/t=12.5.

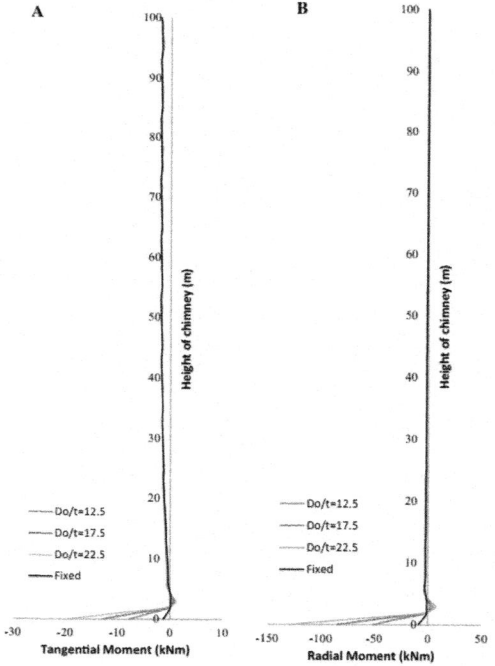

Figure12: aTangential moment,bradial moment of 100m chimney (H/D_b=12) with different D_o/t ratios of raft.

For otherD_o/tratios of 17.5 and 22.5, the tangential moment of chimney with flexible base is increased by 2–6 and 2–8 times, respectively, of that of chimney with fixed base. Significant variation in bending moments in the chimney is found due to the effect of stiffness of raft of piled raft foundation and the effect is less in very tall chimneys. The variation of base moment of 400m chimney is less than that of 100m chimneys due to variation inD_o/tratios. All the above comparisons correspond to the chimney–foundation system supported on loose sand.

EFFECT OF SLENDERNESS RATIO OF CHIMNEY

The effect of slenderness ratio of chimney was studied by considering threeH/D_bratios (H/D_b=7, 12 and 17) of chimney representing the range of stocky to slender chimneys. For stocky chimneys the across-wind loading produces maximum base moment but for other chimneys, along-wind load produces the maximum base moment. It is noted that the along-wind load in chimney withH/D_b=12 is lesser than across-wind load in chimney withH/D_b=7. The chimneys were analysed for these loads causing maximum base moment as considered in design office. The tip deflection of chimney is maximum for chimneys ofH/D_b=17 compared to other twoH/D_bratios. The tip deflection is found less for chimneys ofH/D_b=12 for 100, 200 and 300m chimneys. This is due to the lesser intensity of along-wind loading. For the case of 400m chimney, the tip deflection increases with slenderness of chimney. Therefore, the chimney ofH/D_b=7 subjected to across-wind load shows more tip deflection compared to that ofH/D_b=12 subjected to along-wind load. It is seen that all response of chimneys withH/D_b=12 subjected to along-wind loading is lesser than that in chimneys withH/D_b=7 subjected to across-wind load due to the difference in intensity of loading.

The representative figures for tangential moments in the 100-m chimney for differentH/D_bratios of chimney–piled raft–soil system are shown in Fig.8. From the SSI analysis, it is observed that stocky chimneys (H/D_b=7) with flexible base shows higher tangential moments at a depth of$H/3$ from the top of chimney, especially when the chimney–piled raft system rests on dense sand and rock. This is similar to the response of chimney with fixed base and is different from the response of chimneys withH/D_b=12 andH/D_b=17 with flexible base. The SSI analysis of chimney withH/D_bratios of 12 and 17 shows that the tangential moment is high at the base. The maximum tangential moment in stocky chimney with flexible base is 2.5 times of that in chimney with fixed base. ForH/D_bratios of 12 and 17, it is more than 2.5–8 times of that with fixed base. Similar variation is found in maximum radial moment also. It is found that the effect of SSI is more for slender chimneys (H/D_b=17).

The variation of base moment of 100m chimneys among different slenderness ratios is considerable than that in other chimneys. In general, the SSI effect is more in slender chimneys compared to stocky chimneys because of the less gravitational force

EFFECT OF HEIGHT OF CHIMNEY

The effect of height of chimney was investigated by considering four different heights (H=100, 200, 300 and 400) of chimney. The tip deflection of chimney increases with increase in chimney elevation. Generally, it is seen that the variation of tangential moments in higher elevation chimneys of 300 and 400m (H/D_b=17 and soil type S1) is high when compared to that in chimneys of 200 and 100m. The variation of radial moment is significant in chimneys of all elevations. The variation in base moment of chimney is more than 75% even with the effect of interaction with supporting rock base as compared to fixed base.

SIMPLIFIED EQUIVALENT FIXED BASE MODELS FOR SSI

There is a high demand for developing simplified SSI models used to determine the actual response of chimneys under the flexibility of soil without conducting the rigorous and time-consuming numerical analysis. From the extensive parametric study, simplified equivalent fixed base models were developed using multi-linear regression analysis which will be helpful for practical purposes to evaluate wind response of chimneys considering detrimental effects of SSI. From the extensive SSI analysis, it is found that the base moment of the chimney with soil as flexible base is less than that in the conventional analysis in which fixity at the base of chimney is assumed. Therefore, the reduction of base moment of chimney due to SSI is deemed to be conservative and could be ignored in the design procedure contributing to safer design. However, amplification of tip deflection of chimney and maximum tangential and radial moment in the chimney shell due to SSI have detrimental effects on performance and safety of chimney–foundation system and must be taken into account in any design procedure. The ratio of maximum elastic response of chimney with flexible base to fixed base was derived. This ratio is called modification factor (X).

$$\frac{\Delta'}{\Delta} = X_{(\text{tip deflection})},$$

$$(14)$$

where Δ' is the tip deflection of chimney with flexible base and Δ is the tip deflection of chimney with fixed base.

$$X_{(\text{tip deflection})} = 7.195 - 0.001 \times H + 0.029 \times K_{CS}$$
$$- 1.24 \times K_{RS} - 4.296 \times (\gamma_c / \gamma_s)$$
$$+ 4.97 \times 10 - 9 \times M_{(\text{base})}, \qquad (15)$$

whereHis the height of chimney (m),γ_c is the density of material of chimney–foundation system (kN/m³),$\gamma_s$is the density of soil stratum (kN/m³),$M_{(\text{base})}$is the base moment of chimney with fixed base (kNm) andK_{CS} is the relative stiffness of chimney.

K_{CS}given below is the modified formulation given by Potts and Addenbrooke (1997):

$$K_{CS} = \frac{EI}{E_s \left(\frac{D_b}{2}\right)^4}, \qquad (16)$$

where EIis the bending stiffness of chimney,$D_b$is the diameter at the bottom of the chimney in the plane of deformation and$K_{RS}$is the relative stiffness of raft;K_{RS}given below is the modified formulation given by Fraser and Wardle (1976):

$$K_{RS} = 4E_R(1 - v_S^2)t_R^3 / 3E_S(1 - v_R^2)(D_o - D_i)^3, \qquad (17)$$

where$E_R$is the Young's modulus of raft,$E_s$is the Young's modulus of soil $v_R$is the Poisson's ratio of raft,$v_s$is the Poisson's ratio of soil,$t_R$is the thickness of the raft,$D_o$is the outer diameter of annular raft and$D_i$is the inner diameter of annular raft.

It is seen that the practical range of relative stiffness of chimney is $1<K_{CS}<65$, whereas that of raft is $0.001<K_{RS}<0.5$. The upper limit ofK_{CS}represents chimney resting on loose sand, whereas the lower limit ofK_{CS}represents chimney resting on rock. Hence the lower limit ofK_{CS}points out little interaction effect with soil. The lower limit ofK_{RS}represents a flexible raft resting on rock and the upper limit ofK_{RS}represents a virtually rigid foundation resting on loose sand.

The modification factor for tangential moment in chimney ($X_{(Mt\backslash,\text{chimney})}$) is given as following:

$$\frac{M'_{t(\text{chimney})}}{M_{t(\text{chimney})}} = X_{(M_t\,\text{chimney})}. \qquad (18)$$

where$M'_{t\,(\text{chimney})}$is the maximum tangential moment in chimney with flexible base and$M_{t\,(\text{chimney})}$is the maximum tangential moment in chimney with fixed base.

The equation of modification factor for tangential moment in chimney is given as following:

$$
\begin{aligned}
X_{(M_{t\,chimney})} = {} & -10.502 + 0.005 \times H + 0.087 \times K_{CS} \\
& - 15.544 \times K_{RS} + 9.344 \times (\gamma_c / \gamma_s) \\
& - 1.03 \times 10^{-7} \times M_{(base)}
\end{aligned}
\tag{19}
$$

The modification factor for radial moment in chimney $(X_{(Mr\setminus,\ chimney)})$ is given as following:

$$
\frac{M'_{r(chimney)}}{M_{r(chimney)}} = X_{(M_{r\,chimney})},
\tag{20}
$$

where $M'_{r(chimney)}$ is the maximum radial moment in chimney with flexible base and $M_{r(chimney)}$ is the maximum radial moment in chimney with fixed base.

The equation of modification factor for radial moment in chimney is given as following:

$$
\begin{aligned}
X_{(M_{r\,chimney})} = {} & -16.201 - 0.004 \times H + 0.118 \times K_{CS} \\
& - 21.281 \times K_{RS} + 15.904 \times (\gamma_c / \gamma_s) \\
& - 5.91 \times 10^{-8} \times M_{(base)}
\end{aligned}
\tag{21}
$$

Equations(15), (19) and (21) are valid only for $0.01 < K_{RS} < 0.456$. The effect of SSI is found significant in this range of relative raft stiffness.

CONCLUSIONS

The effect of SSI was investigated for reinforced concrete chimneys with piled raft foundation founded on four different types of soil subjected to wind loads. Piled raft foundations having different thickness of raft, different chimney elevations and different slenderness ratio of chimney were selected for the parametric study. Three-dimensional finite element analysis of integrated soil–foundation–chimney system was carried out. 180 numbers of finite element models were generated. The responses of chimney in terms of lateral deflection, tangential and radial bending moment and base moment were evaluated for the chimney–piled raft–soil system and compared with chimneys of fixed base. The percentage variation was computed for maximum values of moments in chimney obtained through SSI analysis and from fixed base analysis of chimney.

The responses in chimney such as lateral deflection, tangential moment and radial moment increases with increase in the flexibility of soil whereas the base moment of chimney increases with increase in stiffness of the soil. It is also found that the different responses in chimney are increased drastically with decrease in the thickness of raft of piled raft foundation. The variations of tangential and radial moments are higher in slender chimneys. A higher variation of tangential moments is seen in the chimneys of higher elevations whereas the variation of radial moments is significant for all chimneys under consideration.

The following general observations are drawn from the SSI analysis of chimney with piled raft foundation:

- The base moment of chimney is reduced more than 75% from that of chimney with fixed base due to the effect of SSI.

- For stocky chimneys, the maximum tangential moment is found at the base due to SSI whereas it occurs at a height ofH/3m from the top of the chimney for fixed base.

- The effect of SSI is significant only up to a height ofH/40 from the base of the chimney in radial moment variation andH/20 from the base in the case of tangential moment variation.

- The maximum tangential moment in stocky chimney founded on loose sand is increased by 2.5 times of that in chimney with fixed base whereas for slender chimneys it is increased up to eight times of that with fixed base.

- The variation in maximum tangential moment of chimney is double for a chimney with thin raft as compared to that with thick raft when founded on loose sand. Similar variation occurs in radial moment also.

It is concluded that the estimation of the response of slender chimneys due to SSI is very important. Construction of chimneys even with piled raft foundation having skin friction piles in loose sand is not recommended. The effect of SSI is predominant at the base of chimney. Hence, the response variation at base for a distance of at least 1/40th of the height of chimney should be considered for safe design. The present SSI study would be helpful to the design engineers for the optimum selection of geometrical parameters of chimney and foundation.

REFERENCES

1. ACI (307-2008) (2008) Code requirements for reinforced concrete chimneys (ACI 307-08) and commentary. American Concrete Institute, Farmington Hills

2.   Arunachalam S, Govindaraju SP, Lakshmanan N, Appa Rao TVSR (2001) Across-wind aerodynamic parameters of tall chimneys with circular cross section—a new empirical model. Eng Struct 23:502–520

3.   Arya AS, Paul DK (1977) Earthquake response of tall chimneys. In: Proceedings of the sixth world conference, Delhi, pp 1247–1259

4.   Bowles JE (1997) Foundation analysis and design. McGraw-Hill, Singapore

5.   Brown PT (1969) Numerical analysis of uniformly loaded circular rafts on deep elastic foundations. Geotechnique 19:399–404

6.   Cakir T (2013) Evaluation of the effect of earthquake frequency content on seismic behaviour of cantilever retaining wall including soil–structure interaction. Soil Dyn Earthq Eng 45:96–111

7.   Chaudhary MTA (2007) FEM modelling of a large piled raft for settlement control in weak rock. Eng Struct 29:2901–2907MathSciNet

8.   Chowdhury I, Dasgupta SP (2009) Dynamics of structure and foundation—a unified approach. CRC Press, The Netherlands

9.   Chu KH, Afandi OF (1966) Analysis of circular and annular slabs for chimney foundations. J ACI 63:1425–1447

10.   CICIND (2001) Model code for concrete chimneys, Part A. In: The Shell, 2nd edn (revision 1)

11.   Davenport AG (1967) Gust loading factors. J Struct Eng ASCE 93:1295–1313

12.   Davenport AG (1995) How can we simplify and generalize wind loads. J Wind Eng Ind Aerodyn 54(55):657–669

13.   Dewaikar DM, Patil PA (2006) Analysis of a laterally loaded pile in cohesionless soil under static and cyclic loading. Ind Geotech J 36(2):181–189

14.   Fang HY (1991) Foundation engineering handbook. Van Nostrand Reinhold, New York

15.   Fraser RA, Wardle LJ (1976) Numerical analysis of rectangular rafts on layered foundations. Geotechnique 26(4):613–630

16.   Ghosh S, Wilson EL (1969) Dynamic stress analysis of axi-symmetric structures under arbitrary loading. In: Report No. EERC 69-10. University of California, Berkeley

17.   IS: 11089-1984 (reaffirmed 2002) Code of practice for design and construction of ring foundation. Bureau of Indian Standards, New Delhi

18.   IS: 4998(Part 1)-1992 (reaffirmed 2003) Criteria for the design of

reinforced concrete chimneys. Bureau of Indian Standards, New Delhi

19. IS: 875(Part 3)-1987 (reaffirmed 2003) Code of practice for design loads (other than earthquake) for building and structures. Bureau of Indian Standards, New Delhi

20. Jayalekshmi BR, Menon D, Prasad AM (2011) Effect of soil–structure interaction on along-wind response of tall chimneys. IACMAG 2:846–851

21. Koten HV (2005) Wind actions. In: Zurich (ed) The CICIND chimney book: industrial chimneys of concrete or steel. CICIND, Zurich, pp 99–114

22. Kwok KCS, Melbourne WH (1981) Wind induced lock in excitation of tall structures. J Struct Eng ASCE 107(1):57–72

23. Manohar SN (1985) Tall chimneys-design and construction. Tata MacGraw-Hill, New Delhi

24. Mehta D, Gandhi NJ (2008) Time study response of tall chimneys under the effect of soil structure interaction and long period earthquake impulse. In: The 14th world conference on earthquake engineering, China

25. Melbourne WH (1997) Predicting the cross-wind response of masts and structural members. J Wind Eng Ind Aerodyn 69(71):91–103

26. Melerski ES (1990) Simple computer analysis of circular rafts under various axisymmetric loading and elastic foundation conditions. Proc Inst Civ Eng Part 2(89):407–431

27. Menon D, Rao PS (1997a) Estimation of along-wind moments in RC chimneys. Eng Struct 19(1):71–78

28. Menon D, Rao PS (1997b) Uncertainties in codal recommendations for across-wind analysis of RC chimneys. J Wind Eng Ind Aerodyn 72:455–468

29. NEHRP (1994) Recommended provisions for seismic regulations of new buildings. Part 1—Provisions. FEMA 222A

30. Potts D, Addenbrooke T (1997) A structure's influence on tunnelling-induced ground movements. Proc Inst Civ Eng Geotech Eng 125:109–125

31. Pour NS, Chowdhury I (2008) Dynamic soil structure interaction analysis of tall multi-flue chimneys under aerodynamic and seismic force. In: The 12th international conference on IACMAG, India, pp 2696–2703

32. Rajasankar J, Iyer NR, Swamy BY, Goplalakrishnan N, Chellapandi P (2007) SSI analysis of a massive concrete structure based on a novel convolution/deconvolution technique. Sadhana 32:215–234

33. Sáez E, Caballero FL, Razavi AMF (2011) Effect of the inelastic dynamic soil–structure interaction on the seismic vulnerability assessment. Struct Saf 33:51–63

34. Simiu E (1976) Equivalent static wind loads for tall building design. J Struct Eng ASCE 102:719–737

35. Solari G (1982) Along-wind response estimation: closed form solution. J Struct Eng ASCE 108:225–234

36. Tabatabaiefar HR, Massumi A (2010) A simplified method to determine seismic responses of reinforced concrete moment resisting building frames under influence of soil–structure interaction. Soil Dyn Earthq Eng 30:1259–1267

37. Turner T (2005) Industrial chimney foundations. In: Zurich (ed) The CICIND chimney book: industrial chimneys of concrete or steel. CICIND, Zurich, pp 79–98

38. Vickery BJ, Clark AW (1972) Lift or across-wind response of tapered stacks. J Struct Eng ASCE 98(ST 1):1–20

39. Wolf JP (1985) Dynamic soil–structure interaction. Prentice-Hall, New York

Chapter 9

ASSESSMENT OF PILE RESPONSE DUE TO DEEP EXCAVATION IN CLOSE PROXIMITY—A CASE STUDY BASED ON DTL3 TAMPINES WEST STATION

C.G. Chinnaswamy and David N.G. Chew Chiat

Meinhardt Infrastructure Pte Ltd, Singapore, Singapore

ABSTRACT

Ground movements during deep excavations and tunnelling, especially in urban areas, may potentially have major impact on adjacent buildings, structures and utilities. This impact on buildings and structures needs to be assessed by considering the horizontal and vertical displacements induced by deep excavations to determine the necessary mitigation measures. One major factor affecting the degree of severity the impact due to deep excavation may have on the buildings and structures is the type of foundation systems. While methodology in determining the damage category for the buildings on shallow foundation has been quite well established, the methodology for assessing the impact on the pile foundation is not straightforward due to the geometry and complexity of soil structure interaction. Often simplified two-dimensional (2D) or comprehensive three-dimensional (3D) finite element analyses would be carried out for the stage excavation to predict the displacement and stresses in the piles. Suitable protective and preventive measures would need to be designed and implemented for the existing buildings/structures if the damage category falls within the unacceptable range. This paper discusses the analysis and methodology to assess the effect on the pile foundation of a high-rise building due to the deep excavation of the Down Town Line Stage 3 (DTL3) Tampines West (TPW) Station. The approach to assess the geotechnical capacity of the pile as a result of the deep excavation is presented in this paper. Based on the assessment of pile response, predicted movement, structural and geotechnical capacities of the pile, it was found to be within the acceptable limit and the pile foundation has adequate factor of safety with the deep excavation in close proximity.

INTRODUCTION

Ground movements during deep excavations and tunnelling, especially in urban areas, may potentially have major impact on adjacent buildings, structures and utilities. Figure 1 shows the crack on the external walls and columns of the buildings adjacent to deep excavation projects. Hence, it is critical to assess this impact on buildings and structures by considering the horizontal and vertical displacements induced by deep excavations to determine the necessary mitigation measures.

Figure 1: Cracks on external walls and columns of buildings caused by deep excavation.

One major factor affecting the degree of severity the impact due to deep excavation may have on the buildings and structures is the type of foundation systems. While methodology in determining the damage category for the buildings on shallow foundation has been quite well established, the methodology for assessing the impact on the pile foundation is not straightforward due to the geometry and complexity of soil structure interaction. Often, simplified two-dimensional (2D) or comprehensive three-dimensional (3D) finite element analyses would be carried out for the stage excavation to predict the displacement and stresses in the piles. Suitable protective and preventive measures would need to be designed and implemented for the existing buildings/structures if the damage category falls within the unacceptable range.

This paper discusses the case study of analysis and methodology to assess the effect on the pile foundation of a high-rise building due to the deep excavation of the Down Town Line Stage 3 (DTL3) Tampines West (TPW) Station. Figure 2 shows the location map of the TPW station in relation to the DTL3 alignment. It is located in the eastern part of Singapore. TPW Station is located in close proximity to the existing HDB Blocks as shown in Figure 3. TPW Station is a three-level Civil Defence underground station of about 160 m length, 40 m maximum width and 22.6 m deep. The geological formation at the site is mainly old alluvium (OA) soil with overlying fill. The engineering properties of OA material were comprehensively described by Wong et al. (2001), Chiam et al. (2003) and Chu, Goh, Pek, and Wong (2003). Table 1 shows the summary table of the design parameters for the soils. Figure 4 shows the geological profile along the TPW Station.

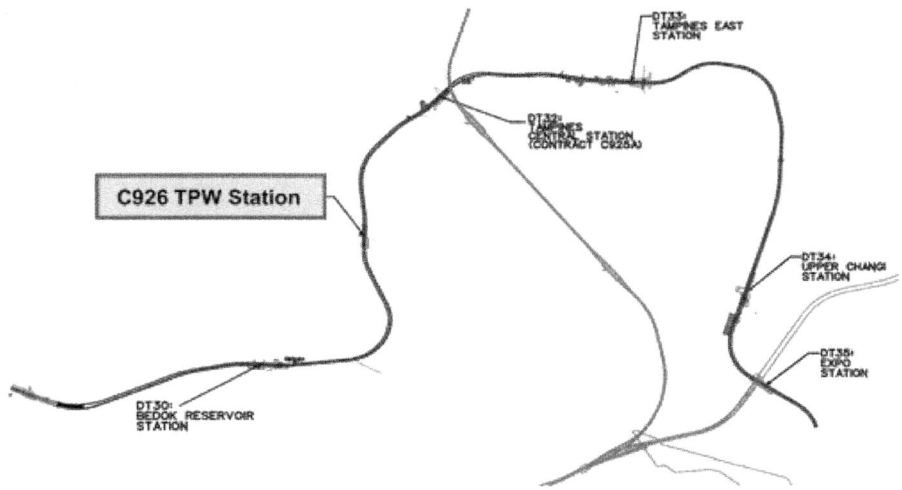

Figure 2: Location map for DTL3 C926 TPW Station.

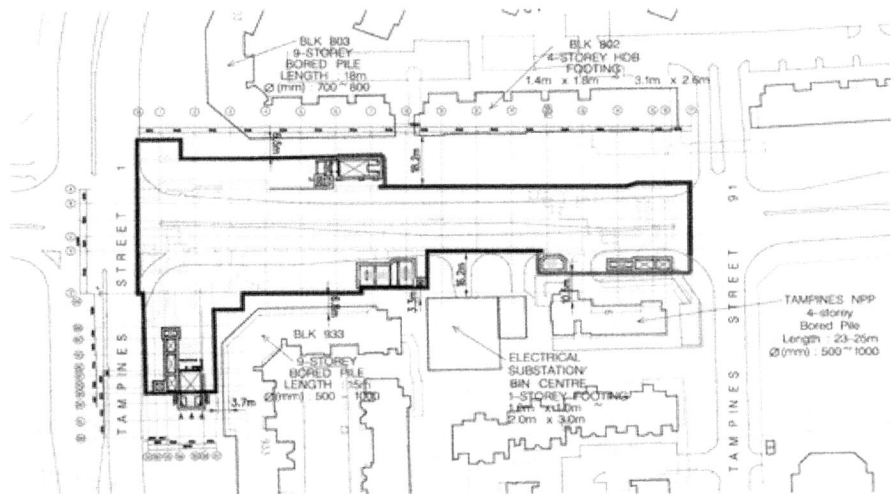

Figure 3: Plan showing proximity of the TPW Station to the HDB Blocks.

Table 1: Summary table of soil parameters

Material	Unit weight (kN/m³)	Strength parameters			Undrained modulus, E_u (MN/m²)	Drained Modulus, E' (MN/m²)	Coefficient of earth pressure at-rest, K_o	Permeability (m/s)
		Total stress S_u (kN/m²)	Effective stress c' (kN/m²)	φ' (°)				
Fill	20	30	0	-	-	8.7	0.5	10^{-7}
E	15	$0.75z + 16.25$ $(20 \leq S_u \leq 35)$	0	15	0.25_u	$E_u/1.2$	1.0	10^{-9}
F1	20.5	-	0	30	-	8.7	0.7	10^{-6}
F2	19	$1.5z + 12.5$ $(20 \leq S_u \leq 50)$	5	25	0.25_u	$E_u/1.2$	1.0	10^{-7}
M	16	$1.285z + 3.575$ for $10 \leq S_u \leq 55$	0	22	0.35_u	$E_u/1.2$	1.0	10^{-9}
Old Alluvium								
OA (E) (N < 10)	20	5 N	0	30	1.0	$E_u/1.2$		10^{-7}
OA (D) (10 ≤ N < 30)	20	5 N	5	32	2 N	$E_u/1.2$		10^{-7}
OA (C) (30 ≤ N < 50)	21	5 N	10	32	2 N	$E_u/1.2$	0.75$^{(b)}$ / 1.0	10^{-7}
OA (B) (50 ≤ N < 100)	21	3 N + 100	10	35	1.2 N + 40	$E_u/1.2$		10^{-7}
OA (A) (N ≥ 100)	21	400	20	35	160	$E_u/1.2$		10^{-7}

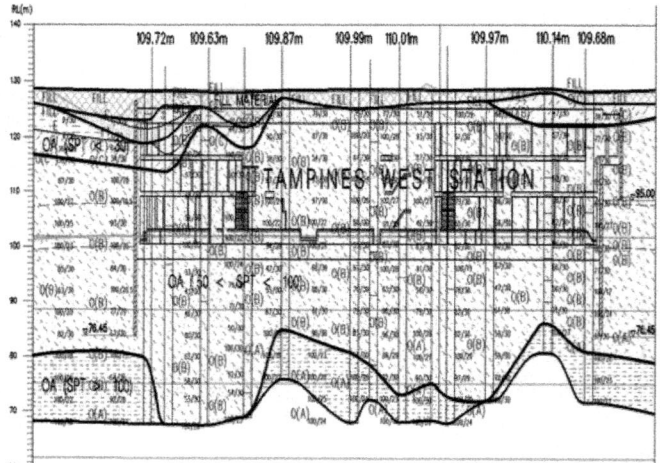

Figure 4: Geological profiles along TPW Station.

Based on geological survey by PWD (1976), the OA is an alluvial deposit that has been variably cemented, often to the extent that it has the strength of a very weak or weak rock. The upper zone of the OA has typically been affected by weathering and has typically penetrated as a discernible front from the surface. All five classes of weathering classification of the OA are encountered at this site.

STAGEWISE DAMAGE ASSESSMENT

Damage assessment of buildings or structures adjacent to deep excavations is a major design consideration in densely built-up areas. These excavations are designed with earth retaining and stabilizing structures (ERSS), which must be robust enough to prevent and minimize any damage to the adjacent structures.

It is necessary to predict the extent of ground movements that may cause damage to the structures. For buildings and structures on pile foundations, the following steps are part of the damage assessment procedure:

1. Predicting the vertical and horizontal movements of building and foundations which are determined from the numerical analysis by considering the foundation contribution in the continuum model or by empirical methods using Gaussian Settlement curve for the case of bore tunnelling works.

2. Damage assessment of the structure based on the predicted vertical and horizontal movements and assuming greenfield conditions with buildings as masonry structures.

3. Study on pile behaviour and response based on the reduction in pile skin friction due to change in stress-field in soil and thus the impact on the geotechnical capacity of the pile. The additional pile displacements, bending moments and shear forces induced on the pile due to the excavation would also be studied.

A case study based on DTL3 C926 TPW Station deep excavation effect on the adjacent pile foundation for a high-rise building is presented in the following sections. In general, the damage assessment procedure as described in Step 2 above for buildings and structures are carried out in three stages, Stages I–III which are discussed in the following paragraphs.

Stage I assessment is a preliminary assessment based on the allowable settlement or rotation according to CIRIA PR 30 (1996). If the predicted settlement contours shows more than 10 mm at the building location or if the settlement gradient is more than 1/500, the building or structure should be subjected to Stage II assessment. The predicted settlement is not only due to ERSS's direct deformation effect, but also the settlement due to ground water draw-down and consolidation settlement in case of clayey soils overlying highly pervious soils. Figure 5 shows the settlement contour around the TPW Station due to deep excavation. HDB Blocks No. 802, 803 and 933 fall within the settlement zone of more than 10 mm. Hence, they are subjected to Stages II and III of the damage assessment procedure.

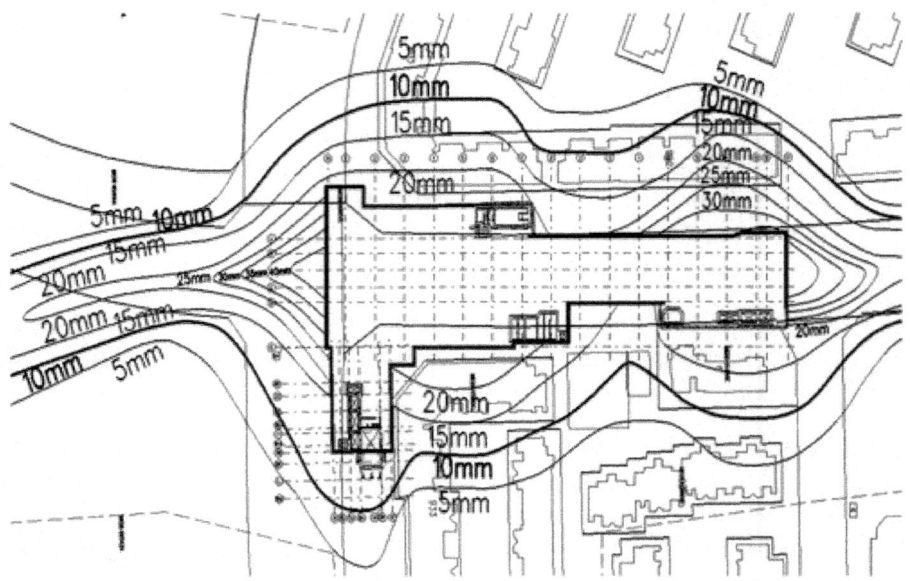

Figure 5: Settlement contour around TPW Station due to deep excavation.

Stage II assessment is based on limiting tensile strain approach adopted by Burland and Wroth (1974), Boscardin and Cording (1989) and Burland (1997, 2008), where the building is idealized as an equivalent deep beam of length, L and height H as shown in Figure 6, with an assumption that the building follows the settlement trough and also the lateral movements induced by deep excavation works as in Figure 7.

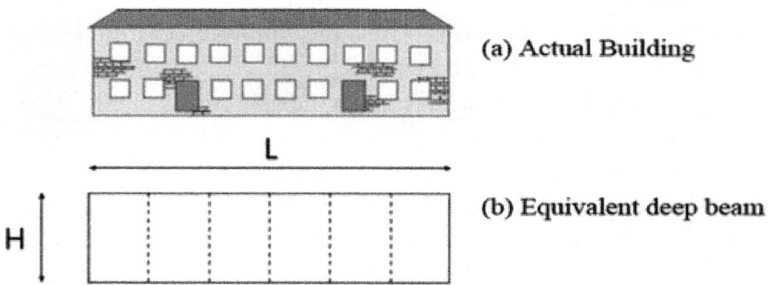

Figure 6: Building idealization for Stage II damage assessment.

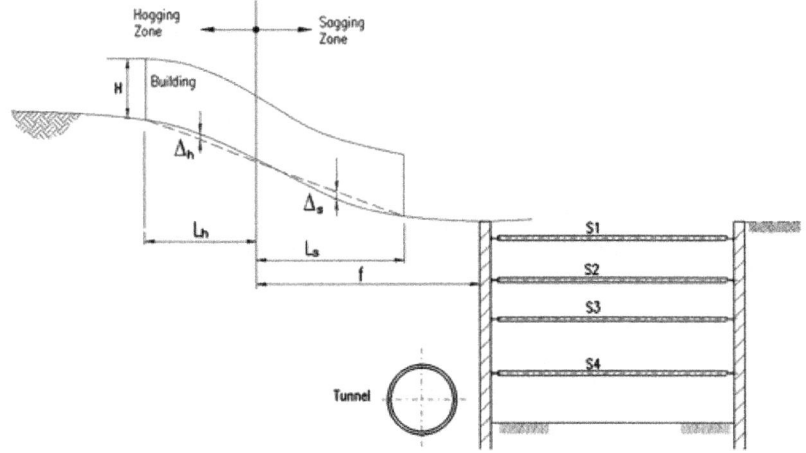

Figure 7: Building deformation—partitioning between sagging and hogging.

The deflection ratio, h/l_h and s/l_s, where suffix "s" is for sagging and "h" is for hogging, is a measure of curvature and various induced strains viz., maximum extreme fibre strain, $b_{(max)}$ (bending), maximum diagonal strain, $d_{(max)}$ (shear) and their respective resultants, ε_{br} and ε_{dr} when combined with horizontal strain, h The procedure to estimate various strains is well described in publications of Burland (2008) and Loganathan (2011).

The maximum induced strain among these resultant strains is set as the limiting tensile strain, ε_{lim} and a range in the limiting tensile strain is used to categorize the damage to the building from being at "Negligible" through to "Very Severe" risk as shown in Table 2. Description of typical damage according to degree of severity with particular reference to ease of repair of plaster and brickwork or masonry can be seen from references by Burland (2008), Civil Design Criteria of LTA (2008), and Loganathan (2011).

Table 2: Relationship between category of damage and limiting tensile strain

Category of damage	Normal degree of severity	Limiting tensile strain, ε_{lim} (%)
0	Negligible	0–0.05
1	Very slight	0.05–0.075
2	Slight	0.075–0.15
3	Moderate	0.15–0.3
4 and 5	Severe to very severe	>0.3

Stage III assessment is a detail assessment of the structures with numerical analyses. In general, all structures that have been classified in the "Moderate" or higher damage risk categories during Stage II assessment are classified as

"Sensitive Structures". However, historical and sensitive structures with a "Slight" damage category and any structures on pile and mixed foundations will also be subjected to Stage III assessment according to LTA Civil Design Criteria (2008). The Stage III assessment can be performed by using either the method proposed by Potts and Addenbrooke (1996) or using numerical modelling by incorporating the building stiffness as well. In addition, all reinforced concrete structures will be assessed based on their service-ability limits. Two-dimensional analyses are usually carried out instead of 3D analyses due to the complexity of the 3D modelling procedures. While selecting the 2D numerical approach for Stage III assessment, the following factors should be kept in view:

- Since piles are discrete elements, smearing of pile stiffness should be suitably considered.

- If as-built information is not suitable like pile length, geophysical survey to ascertain the pile length and a set of parametric study needs to be carried out for a possible range of pile lengths to make it reasonably conservative.

ASSESSMENT OF IMPACT ON PILE FOUNDATION BY NUMERICAL ANALYSES

This section will discuss the numerical analyses aimed to check for the changes in geotechnical pile capacities which are quite likely to happen due to the settlements and changes in stress field in the soil surrounding the piles due to wall deflection and base heave during excavation. This stress field changes will lead to a reduction in the effective normal stress and thus the skin friction on the piles, which in turn will increase in end bearing pressure. These changes need to be checked for the ultimate end bearing capacity of the pile. The typical section of ERSS for TPW station adjacent to the HDB Block 803 is as shown in Figure 8.

For a typical case of piles supporting 9-storey HDB Blocks No. 803 and 933 adjacent to deep excavation for TPW Station, the finite element model is shown in Figure 9. This model is a 2D plane strain model with 15 noded isoparametric triangular elements for soil layers and 5 noded plate elements for all structural elements except for struts for which node-to-node anchor elements/plane truss element were adopted. The piles were also modelled as plate elements with their stiffnesses smeared for the average pile spacing and the soil–pile interaction was modelled by using the interface elements around the pile elements which were assigned with reduced soil strength properties (decreased by a strength reduction factor, R_f). In the stress analysis, both x and y displacements were set to zero at the bottom boundary, whereas at the truncated sides, only nodal displacements in the x-direction were set to zero.

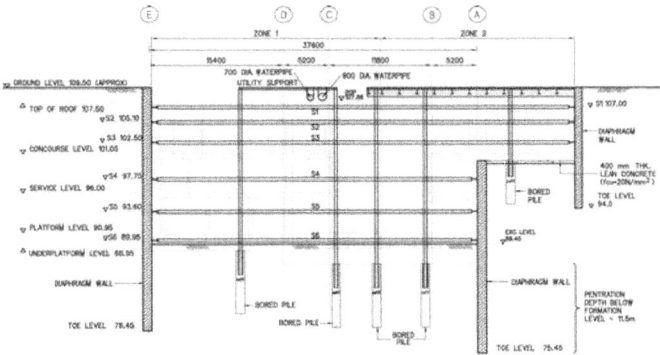

Figure 8: Typical sections of ERSS for TPW station adjacent to the HDB Block 803.

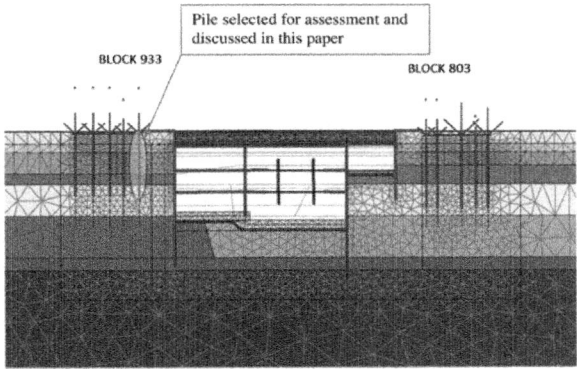

Figure 9: Finite element meshes showing the pile foundation arrangement adjacent to a deep excavation (encircled closest pile considered in this study).

The soil constitutive model used in the analysis was the Mohr–Coulomb model, an elastic perfectly plastic bilinear stress strain model. Undrained behaviour was set for all clay layers by choosing undrained material type, undrained elastic modulus, E_u, and undrained strength parameters of S_u and $\Phi_u = 0$ and drained behaviour was set with drained material type, effective elastic modulus, E' and effective stress parameters viz, c' and φ' for all sandy soils. However, since OA soil behaviour is in the transition between drained and undrained behaviour, both these cases were considered thus leading to two sets of numerical analyses covering drained and undrained cases. In all cases of numerical analyses, pore water pressure of soil elements was determined by a steady-state 2D seepage analysis which was carried out prior to elasto–plastic stress analysis for each excavation/construction stage. Calculated pore water pressures in this way were used for the effective stress calculations and then used in the elasto–plastic analysis.

The pile foundation for the HDB Block No. 803 and 933 consists of 700 mm diameter bored piles at approximately 3.5 m centre-to-centre spacing. The lengths of the piles were not available in the as-built drawings retrieved from BCA. Initial estimate of the piles was based on the estimated working load from the columns. Later, geophysical survey was carried out to verify the estimated pile length. Finally, pile length of 18 m was adopted in the analyses and study. For ERSS design, Sandi, Shen, Leung, Liew, and Kho (2007) compared 2D and 3D FE analyses with field measurements and a range of smearing factors to be adopted in 2D FE analysis and concluded that using smearing factor of $3d_{pile}$ gives similar predictions using 3D numerical analysis. In 2D analyses, we adopt a smearing effect for the discrete structural elements like piles to simulate the discrete element using 2D plane strain analysis.

RESULTS OF NUMERICAL ANALYSES

Figures 10 and 11 show the vertical and horizontal soil displacements, respectively, for the complete cycle of excavation and backfill. As shown in Figure 10, the downward displacement on both sides of the plate element, which represents the pile, is unsymmetrical. The vertical downward displacement on the side nearer to the excavation is larger than the vertical downward displacement on the other side further away from the excavation. This has resulted in the unsymmetrical changes in the shear stress on both sides of the plate element. This will be discussed in more detail in Section 5 of this paper. On the other hand, as shown in Figure 11, the horizontal displacement of the soil on the side of the plate element is almost similar. This is due to the reason that the pile element has been modelled as a thin plate element and thus has not caused any significant change in the stress field and vertical movement of the soil on both sides of the plate element. This has resulted in symmetrical decrease in the normal stress on the plate element.

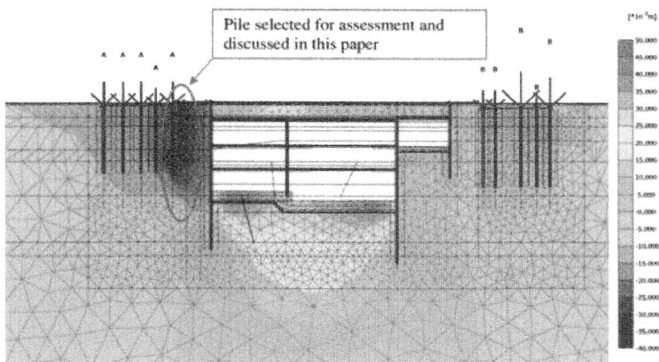

Figure 10: Vertical soil displacements due to the complete cycle of stage excavation and construction of the station structure.

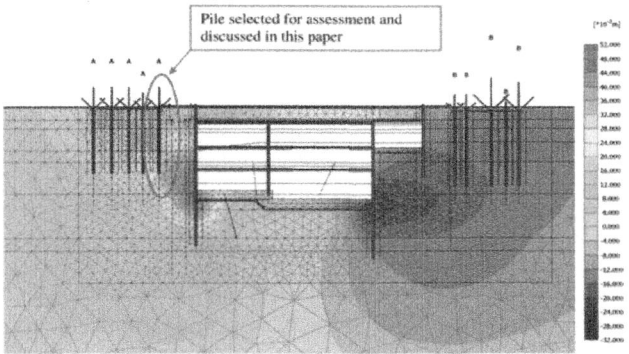

Figure 11: Horizontal soil displacements due to the complete cycle of stage excavation and construction of the station structure.

REDUCTION IN SKIN FRICTION RESISTANCE OF THE PILE

For the pile close to the deep excavation, comparison of effective normal stress distribution on the pile shaft before and after excavation is shown in Figure 12. The kinks especially at elevations closer to the ground levels seen at the plot of effective normal stress after excavation are due to the preloading effect. The resultant of the normal stress on pile interfaces both on soil side and excavation side is same. However, the resultant of the normal stress before the excavation is 1,155 kN/m, whereas after excavation, it reduces to 878 kN/m, which is about 24% reduction from the value before excavation. The response of pile in terms of the normal stress acting on the pile is almost symmetrical due to the reason that the pile has been modelled as a plate element with smeared properties of the pile. This thin plate element has caused insignificant change in the stress field on both sides of the plate due to the low stiffness. In addition, the results also indicate that there is no build up of active and passive soil pressure on the two sides of plate element when soil movement occurred. This is likely due to the low stiffness of the plate element.

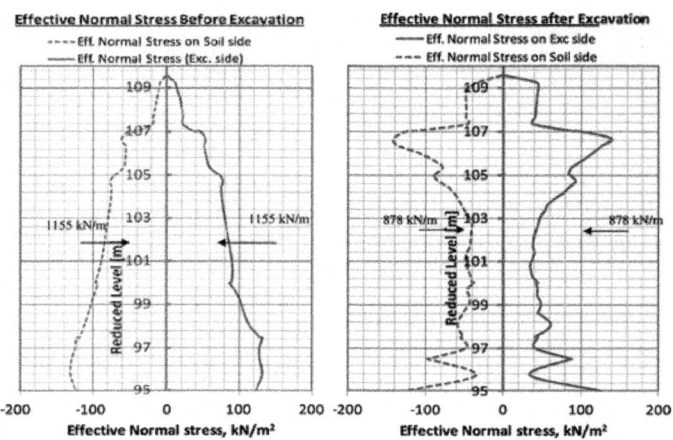

Figure 12: Effective normal stress distribution on the pile closest to the excavation before and after excavation.

In the conventional method of checking the geotechnical capacity of bored pile, empirical relationship, fs = Ks × N is commonly used to determine the ultimate skin friction, where N = SPT values and the coefficient, Ks, varies from 1.5 to 2.5 for stiff-to-hard cohesive soils, including Bukit Timah Granite and Jurong formation soils and 2–3 for dense and hard, cemented OA soils. For both the cases, the limiting values of fs are specified CP04:200 as 150 and 300 kPa respectively. Similar way of estimating the ultimate end bearing from the SPT-N values is also described in CP 04:2003 (2003). While adopting this empirical method for estimating the pile's ultimate geotechnical capacities, the same reduction factor for the skin friction due to adjacent deep excavation can be applied in order to include the effect of adjacent deep excavation on ultimate skin friction of the piles.

Comparison of developed skin friction on the both sides of the plate elements which simulate the pile shaft before and after excavation is shown in Figure 13. The sign convention for the representation of the graphs is positive for upward direction shear stress and negative for downward direction shear stress for the face near to excavation and is negative for upward direction shear stress and positive for downward direction shear stress for the face away from excavation. The shear stress distribution on the two sides of the plate element of the pile before excavation is both showing upward direction but is unsymmetrical. The unsymmetrical distribution is due to the reason that the pile is on the edge of the series of piles for the building.

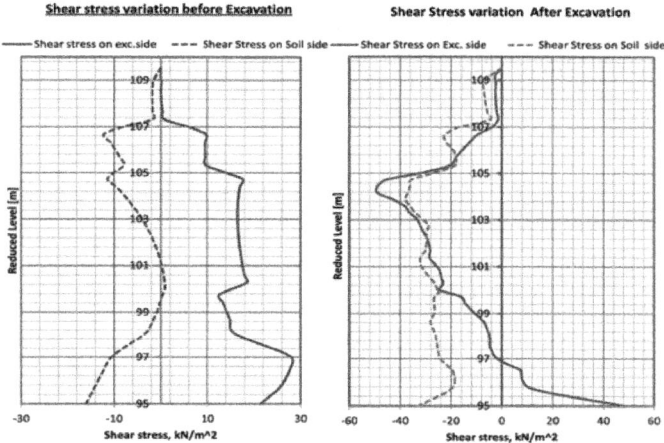

Figure 13: Comparison of development of skin friction on the pile closest to the excavation before and after excavation.

After excavation, there is soil movement towards the excavation and also subsurface soil settlement. The resulting soil movement due to the excavation is such that there is also similar change in the horizontal displacement on both sides of the plate element and there is more downward displacement of the soil on the side of the plate element closer to excavation than the other face away from excavation. As a result of this soil movement, there is an increase in the upward shear stress on the face of the plate element away from the excavation, while there is a decrease in the upward shear stress and increase in downward shear stress on the face of the plate element near to the excavation. This is likely to be caused by the settlement of the subsoil layers and induces negative skin friction on the pile. The effect of the change in shear stress acting on the pile on its end bearing pressure and overall geotechnical capacity is discussed in more detail in Sections 6 and 7.

INCREASE IN END BEARING PRESSURE OF THE PILE

In order to maintain the force equilibrium condition, the reduction in skin friction will obviously lead to increase in pile axial force and thus the pile end bearing pressure. Figure 14 shows the pile axial force variation with depth. Before excavation, the axial force is 112 kN/m at the toe of the plate element.

After excavation, the axial force at the toe of the plate element increases to 275 kN/m due to the increase in down drag of the soil as a result of settlement. Assuming that the displacements and stress field changes in the soil due to deep excavation would not change the bearing capacity of the soil, it is necessary to check whether this greater end bearing pressure is within the allowable end bearing capacity of the soil layer at the pile toe level. The assessment of pile geotechnical capacity will be described in Section 7.

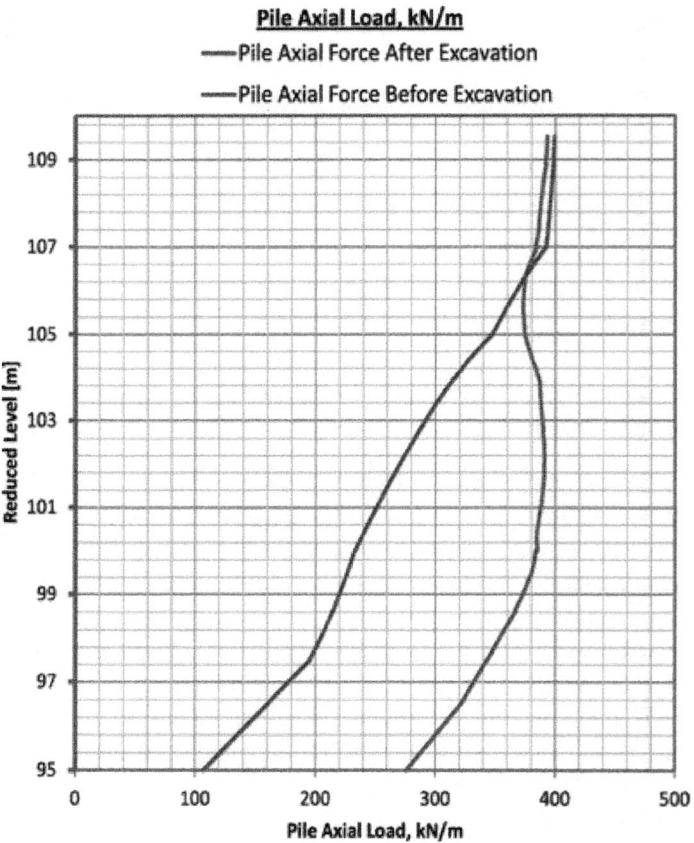

Figure 14: Variation of pile axial force with depth at the pile closest to the retaining wall before and after excavation.

ASSESSMENT OF PILE STRUCTURAL AND GEOTECHNICAL CAPACITY

For the pile structural capacity, it has been checked using the bending moment and shear force diagram obtained from the analyses and multiplied with the smearing factor and found to be acceptable. The details for the structural capacity check is not included in this paper as it is a straightforward procedure and commonly used by engineers. This section will focus on the assessment of geotechnical capacity of the pile as a result of the ground movement caused by deep excavation of TPW Station, which is important to ensure the safety of the building but not commonly checked by engineers probably due to the difficulties in interpreting the results from the analyses and understanding the response of the pile before and after excavation. Table 3 shows the comparison of the pile shaft friction and factor of safety for the cases before and after excavation. The total mobilized skin friction per pile is calculated by multiplying the total shear stresses along the plate element on both faces by the smearing factor. The total mobilized skin friction was reduced from 1,004.5 to 490 kN. The average skin friction is obtained by dividing the total skin friction by the shaft area. The average skin friction is reduced from 30.6 to 15 kPa. After the excavation, the soil has been disturbed along the pile. The load transfer from the pile to soil will be redistributed. The reduction in the shaft resistance will be compensated by the increase in the end bearing resistance. Hence, it is not necessary to evaluate the factor of safety of the pile in terms of shaft friction.

Table 3: Summary table of comparison of pile shaft friction and factor of safety for the cases before and after excavation

Developed skin friction	Before excavation	After excavation
Skin friction mobilized	287 kN/m	140 kN/m
Skin friction per pile	287 × 3.5 kN	140 × 3.5 kN
	=1,004.5 kN	=490 kN
Average skin friction pressure	1,004.5/shaft area	490/shaft area
	=1,004.5/32.8	=490/32.8
	=30.6 kPa	=15 kPa
	Ult skin friction (2.5 N) = 125–250 kPa (Average *SPT* values are in the range of 50–100)	Shaft friction will be disturbed after excavation

Table 4 shows the comparison of the pile end bearing and factor of safety for the cases before and after excavation. The end bearing force was obtained from the axial force at the toe of the plate element multiplied with the smearing factor. The end bearing force increases from 420 kN per pile before excavation to 963 kN per pile. This is corresponding to the increase of end bearing pressure on the soil from 1,010 to 2,502 kPa. The end bearing pressure is obtained by dividing the end bearing force per pile by the pile base area. The SPT-N value for the soil at the toe level of the pile is approximately 80. Based on theSPT-N value of 80 and limiting end bearing pressure, Fb = 60 N, the limiting end bearing pressure is 4,800 kPa. Thus, the factor of safety of the pile in terms of end bearing is adequate.

Table 4: Summary table of comparison of pile shaft friction and factor of safety for the cases before and after excavation

Developed end bearing pressure	Before excavation	After excavation
End bearing force/m	112 kN/m	275 kN/m
End bearing forces/pile	112 × 3.5 = 420 kN	275 × 3.5 = 963 kN
Developed end bearing pressure	392/pile area = 1,018 kPa	963/pile area = 2,502 kPa
Ultimate end bearing	Based on $N = 80$ and $Fb = 60 N$	Based on $N = 80$ and $Fb = 60 N$
	$Fb = 4,800$ kPa	$Fb = 4,800$ ka
FOS (end bearing)	4.7 (minimum and based on $N = 80$)	1.9 (minimum and based on $N = 80$)

The above assessment for the geotechnical capacity of the pile has considered the increase in end bearing pressure of the piles predicted as by the analyses, and has calculated the factor of safety in terms of end bearing for the piles as a result of the station excavation. These show that the factor of safety is still sufficient to meet the minimum requirement.

Before the station is constructed, the piles will have a working load W applied at the pile head. This will be almost entirely resisted in skin friction as the OA strata are highly competent and the piles would predominantly be friction piles. The end bearing pressure at the base of the piles will be very small. During station excavation, there will be some relatively small deformations of the ground outside the diaphragm walls. As a result, the piles are likely to settle by a small amount and this will cause some redistribution of the friction and end bearing components of the pile load, with some increase in end bearing and a corresponding reduction in skin friction. However, the ultimate pile capacity has not been reduced. At the end of the station construction, the same working load W is applied to the pile (assuming no redistribution of load between adjacent piles); this is still resisted mainly by skin friction and by an increased amount of end bearing. The pile will have the same ultimate capacity. The only change is that the pile will simply have settled by a small amount, and there will have been a redistribution of the friction and end bearing components resisting the original pile working load.

Table 5 shows the summary table of the assessment of the pile. The pile lengths of 18 m have been considered for Blocks 803 and 933. In view of the uncertainty due to the unavailability of as-built information for the pile foundation, geophysical testing was essential to verify the assumed pile length. As the ground conditions are principally OA, the piles are acting as predominantly friction piles, with little or no load acting on the pile base. This means that the pile was mainly resisted by friction and has little resistance from end bearing in the original state. After excavation, the pile load transfer to soil merely change its path from via shaft friction to via end bearing at toe.

Table 5: Summary table of impact on the pile foundation on HDB Block 803 and 933

Building	Distance from retaining wall (m)	Foundation type	Pile length from ground level (m)	Maximum pile/footing absolute movements (mm)	Maximum horizontal pile relative movement (mm)	Maximum bending moment due to excavation (kN/m)	Differential settlement angular distortion
Block 933 (9 Storey)	7	Bored pile foundation	18 (GPR)	25(Hor)	15	24	1:2100
			15 (Est)	18.6 (Vert)			
Block 803 (9 Storey)	6.5	Bored pile foundation	18 (GPR)	18 (Hor)	5	10.6	<1:10000
			15 (Est)	8 (Vert)			

As shown in Table 5, Blocks 803 and 933 are predicted to settle by 8 and 19 mm, respectively, assuming the stiffness for the OA to correspond to $E_u/N = 3$. These settlement predictions are likely to be conservative because the stiffness of the OA is generally higher than given by $E_u/N = 3$. The inherent stiffness of the buildings will also lead to load redistribution between the piles and lead to smaller settlement than prediction.

CONCLUSIONS AND RECOMMENDATIONS

In conclusion, the damage assessment approach and procedure for buildings on pile foundation due to adjacent deep excavation or tunnelling as discussed in this paper is reasonable. The variations of the effective normal stress on the pile before and after the deep excavations are also examined and then an approach for assessing the effect on the ultimate skin friction and end bearing of the pile is presented. The redistribution of load transfer from the pile to the soil has also been highlighted. The analysis results show the phenomena of load transfer in pile from shaft friction to end bearing during excavation for the proposed MRT Station. The transfer of load from shaft friction to end bearing is associated with a small amount of settlement as reflected in the analysis.

The decrease in shaft friction and the increase in end bearing of pile do not compromise the overall capacity of the pile. The only change is the pile will settle by a small amount which has insignificant impact to the existing building. The predicted pile/footing settlement will likely be small, in the range of 5–25 mm in competent OA, which has insignificant impact to the buildings. Adopting $E_u = 3 N$ for prediction of building settlement is reasonable. The piles have adequate FOS despite the transfer of resistance from shaft friction to end bearing.

Based on the assessment of pile response, predicted movement, structural and geotechnical capacities of the pile, it was found to be within the acceptable limit and the pile foundation has adequate factor of safety with the deep excavation in close proximity. Using the proposed method, the complete damage assessment of buildings supported on pile foundations can be carried out. It is also shown that by using smearing factor of $3d_{pile}$, the 2D FEM analyses results are appropriate.

ACKNOWLEDGEMENT

The authors would like to thank Mr Song Siak Keong of Land Transport Authority (LTA) for allowing the information for the project to be published in this paper.

REFERENCES

1.   Boscardin, M. D., & Cording, E. G. (1989). Building response to excavation-induced settlement. Journal of Geotechnical Engineering, 115, 1–21.10.1061/(ASCE)0733-9410(1989)115:1(1)

2.   Burland, J.. (2008, 16 de Diciembre de). The assessment of risk of damage to buildings due to tunneling and excavations. Jornada Tecnica de Movimientos de Edificios Inducidos por Excavaciones, Barcelona.

3.   Burland, J. B. (1997). Assessment of risk of damage to buildings due to tunneling and excavation. In Ishihara (Ed.),Earthquake geotechnical engineering (pp. 1189–1201). Rotterdam: Balkema.

4.   Burland, J. B., & Wroth, C. P. (1974). Settlement of buildings and associated damage, SOA review. In Conference on settlement of structures (pp. 611–654). Cambridge: Pentech Press.

5.   Chiam, S. L., Wong, K. S., Tan, T. S., Ni, Q., Khoo, K. S., & Chu, J. (2003). The old alluvium. In Proceedings underground Singapore 2003 (pp. 409–440). Singapore: Nanyang Technological Singapore.

6.   Chu, J., Goh, P. P., Pek, S. C., & Wong, I. H. (2003). Engineering properties of the old alluvium soil. In: Proceedings underground Singapore 2003 (pp. 285–315). Singapore: Nanyang Technological Singapore.

7.   Civil design criteria—Revision A7 for road & rail transit systems. (2008). Land Transport Authority (PED/DD/K9/106/A6).

8.   CIRIA PR 30. (1996). Prediction and effects of ground movements caused by tunneling in soft ground beneath urban areas (Project Report 30). London: Construction Industry Research and Information Association.

9.   CP 04:2003 (Singapore Standard). (2003). Code of practice for foundations.

10.   Loganathan, N. (2011). An innovative method for assessing tunneling induced risks to adjacent structures. PB2009 William Barclay Parsons Fellowship Monograph 25. New York, NY: Parsons Brinckerhoff.

11.   Potts, D. M., & Addenbrooke, T. I. (1996). The influence of an

existing surface structure on the ground movements due to tunneling. In International Symposium on Geotechnical Aspects of Underground Construction in Soft Ground (pp. 573–578). Rotterdam: A A BALKEMA.

12. PWD. (1976). Geology of the Republic of Singapore. Singapore: Author.

13. Sandi, M. S., Shen, R. F., Leung, C. F., Liew, Y. K., & Kho, C. M. (2007). Comparison of 2D and 3D FEA with measurements of pile response adjacent to deep excavation. Underground Singapore 2007. Singapore.

14. Wong, K. S., Li, W., Shirlaw, J. N., Ong, J. C. W., Wen, D., & Hsu, J. C. W. (2001). Old alluvium: Engineering properties and braced excavation performance. In Proceedings underground Singapore (pp. 210–218). Singapore: Nanyang Technological Singapore.

Chapter 10

INTERPRETATION OF AUGERED CAST IN PLACE PILE CAPACITY USING STATIC LOADING TESTS

A. W. Stuedlein, S. C. Reddy and T. M. Evans

School of Civil and Construction Engineering, Oregon State University, 101 Kearney Hall, Corvallis, OR 97331, USA

ABSTRACT

Instrumented static loading tests present an effective tool for appropriate engineering of piled foundations. Whether instrumented or not, considerable effort can be expended to determine the interpreted failure load developed during the loading test, and the determination of an appropriate capacity is often subject to regulatory review and discussion, which may be complicated by the large number of interpretation methods available and the large range in interpreted capacities that could result. This paper focuses on the differences in the interpreted failure load for augered cast in place (ACIP) piles and seeks to determine which methods are suitable and which methods are inappropriate for the interpretation of ACIP piles. First, a review of various capacity interpretation methods is provided with emphasis on those methods cited in the International Building Code (). Then, an ACIP pile case history database used to illustrate differences in interpreted capacity is described, followed by the presentation of the resulting differences. Recommendations for and against the use of several failure load interpretation methods are made. This paper emphasizes the importance of both the interpretation of the failure load as well as the consequence of a particular methodology on displacement – a critical performance measure.

INTRODUCTION

Augered cast in place (ACIP) piles, also known as continuous flight auger piles, present an economical deep foundation alternative to drilled shaft or driven pile foundations because the borehole is excavated and grouted in one continuous operation. Owing to the use of a pressurized grouting protocol, the unit toe and shaft resistances of ACIP piles are often larger than those of drilled shafts (e.g. Stuedlein *et al.*, 2012), though it is recognized that little toe resistance is mobilized in response to working loads. Effective use of ACIP piles is often achieved using pile loading tests, preferably instrumented, but more commonly un-instrumented. In the cases when ultimate geotechnical resistance is not realized a discussion between the regulatory official (e.g. building official), owner, design consultant and contractor is initiated to determine the "capacity" or "failure load" as judged from the test results. Such discussions can quickly lead to frustration over the method of interpretation, despite the commonly held goal of providing the most advantageous and acceptable amount of "useable" pile resistance. The cost of ignoring significant useable pile resistance, though not quantified herein, can be considerable based on anecdotal evidence observed by, and communicated to, the authors.

Although not the focus of this paper, one source of differences in interpreted capacity may be attributed to experimental limitations. Pile loading tests are expensive to carry out, and as a result, savings are attempted by use of undersized reaction frames and reaction piles, undersized hydraulic jacks and uncalibrated load cells or none whatsoever. On the other hand, excessive design conservatism often leads to the construction of a pile with significantly larger capacity than assumed, resulting in insufficient reaction and test frames. These factors directly result in a smaller probability of realizing geotechnical failure of the foundation element being tested, and thus achieving a sufficient enough displacement to evaluate the "true" ultimate resistance to load provided by the pile. Additionally, variations of loading test protocols from accepted best practices (e.g. the maintained load test, Procedure B in ASTM D1143/D1143M-07, 2013) contribute to poorly defined load–displacement (Q–δ) curves and serve to hinder interpretation efforts. Foundations that do not exhibit plunging behavior present particular challenges, especially for drilled foundations, which often exhibit displacement hardening Q–δ responses (Kulhawy and Hirany, 2009) when designed to bear on a competent soil layer.

Perhaps the most significant source of differences in the interpreted capacity stems from the number of methods available for the analysis of loading test data. For example, Hirany and Kulhawy (1988) identified over 40 different procedures available to provide an interpreted failure load, which ranged from graphical construction techniques to absolute settlement limits to methods based on the rate of settlement, among others. Studies by Fellenius (1975) and van Weele (1982), using eight and 12 interpreted failure load criteria respectively, found that the resulting capacities from the various methods could return a spread of 40 and 100% between the lowest and highest values respectively (Kyfor *et al.*, 1992). Differences in the interpreted failure load for ACIP piles form the focus of this paper; specifically, which method(s) are suitable for the interpretation of capacity of ACIP piles, and which method(s) are inappropriate? First, a review of various capacity interpretation methods is provided with emphasis on those methods cited in the International Building Code (IBC, 2012). Then, an ACIP pile case history database used to illustrate differences in interpreted capacity is described, followed by the presentation of the resulting differences. Recommendations for and against the use of several failure load interpretation methods are made. This paper emphasizes the importance of both the interpretation of the failure load as well as the consequence of a particular methodology on displacement – a critical performance measure.

REVIEW OF SELECTED CAPACITY INTERPRETATION METHODS

Fellenius (1975, 1980) and Hirany and Kulhawy (1988), among others, provide excellent reviews of the available methodologies for the interpretation of the pile failure load from a static pile loading test and the factors that can affect the capacity interpretation. Therefore, for brevity, only selected interpretation methods are addressed herein; a discussion of the appropriateness of the selected methods in the context of the ACIP pile loading test case histories are provided later in this paper. The methods selected herein were based on their specification in the IBC (2012) and historical precedent with regard to drilled foundations. Note that the most appropriate term for the load resulting from the various methodologies is the "interpreted failure load" (Hirany and Kulhawy, 1988; Kyfor *et al.*, 1992), $Q_{f,int}$. The *failure load*, or from the geotechnical perspective, the *ultimate resistance*, occurs during pile plunging and is defined as the continuous displacement of the pile head with no appreciable further increase in the application of pile load, sometimes associated with a displacement softening response. The term "failure load" is inappropriate for non-plunging piles. Additionally, the common term "ultimate capacity" is also inappropriate, the modifier "ultimate" being redundant and should be avoided in any discussion involving foundations (Fellenius, 2012).

Perhaps the most well known failure load in North America is that proposed by Davisson (1972) for driven piles termed the Davisson Offset Limit (DOL) method, accepted by IBC (2012). In this approach, the intersection of an offset elastic shortening line with the measured Q–δ curve of the pile indicates the interpreted failure load (Fig. 1). The offset δ_{off}, equal to the sum of 3.8 mm (0.15 in.) and the pile diameter B_p, divided by 120, were suggested by Davisson

(1972) based on the soil quakes (i.e. the displacement associated with the transition from the assumed elastic to perfectly plastic toe bearing response in wave equation analyses) developed for piles of diameters of approximately 0.3 m driven with an impact hammer. The interpreted failure load for the DOL method $Q_{f,int}$ occurs at the intersection between the load-displacement curve and the offset limit line for a pile head displacement δ_{int}, according to equation (1)

$$\delta_{int} = 3 \cdot 8 \text{ mm} + \frac{B_p}{120} + \frac{Q_{f,int}L_P}{A_pE_p}$$

(1)

where B_p is the pile diameter (mm), L_p is the length of the pile (mm), A_p is the cross-section area of the pile (m²), E_p is the elastic modulus of the pile (kPa) and $Q_{f,int}$ is in kN.

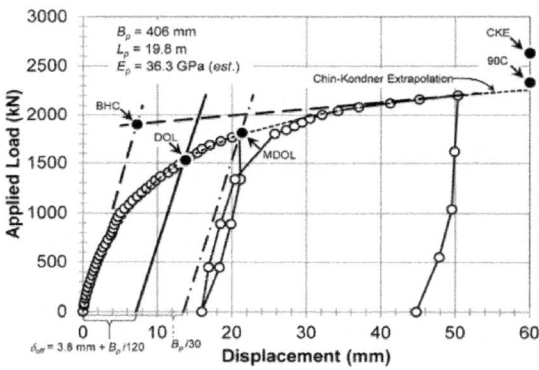

Figure 1: Illustration of selected interpreted failure load methods for 406 mm diameter augered cast in place (ACIP) pile (case 32): test performed in accordance with constant time interval test (Procedure D, ASTM D 1143) with 5 min hold duration.

Owing to the prestressing of soil near the toe of a driven pile, a toe quake on the order of δ_{off}=6 mm for a small diameter driven pile may be sufficient to mobilize considerable toe resistance. However, construction of drilled foundations involves excavation and subsequent casting of concrete in place, resulting in a significantly different stress history (and, therefore, lower stresses) for a soil element near the toe of the foundation, directly contributing to a softer initial slope of the Q–δ curve. It is widely accepted that significantly more toe displacement is required to mobilize substantial toe resistance of drilled foundations compared to driven piles (Vesic, 1977; Hirany and Kulhawy, 1988, 1989; Davisson, 1993; O'Neill and Reese, 1999; Kulhawy and Hirany, 2009). Brown *et al.* (2010) suggested that a diameter normalized toe displacement of 4–5% is required to mobilize significant resistance of drilled foundations; scale effects may therefore be incorporated into the estimation of toe resistance by considering a diameter-dependent toe bearing resistance model. Kyfor *et al.* (1992) suggested the following modification of the DOL method (MDOL)

$$\delta_{int} = \frac{B_p}{30} + \frac{Q_{f,int}L_P}{A_p E_p}$$

(2)

which results in an increase in the offset displacement on the order of 1.75 to two times that of the DOL offset for the range in pile diameters investigated herein. Indeed, Davisson (1993) noted that δ_{off} should be increased by a factor ranging from two to six times that for driven foundations should the DOL method be applied to drilled foundations.

Owing to the softer initial response of the Q–δ curve associated with drilled foundations with slenderness ratios less than 20–25, as shown in Fig. 2, Kulhawy and his colleagues proposed the use of the slope tangent capacity in compression (STC) (Kulhawy, 2004; Kulhawy and Hirany, 2009). The STC approach is essentially similar to the DOL, except that the initial slope of the Q–δ curve is substituted for the slope of the elastic compression line. The initial slope of the Q–δ curve is similar to that of the elastic line for slenderness ratios greater than 20–25 (e.g. Figs. 1 and 2) due to the minimal toe load transfer at small (i.e. less than 5 mm) pile head displacements. Note that the point where the initial slope deviates from the observed Q–δ curve marks the L_1 point associated with L_1–L_2 method proposed by Hirany and Kulhawy (1988).

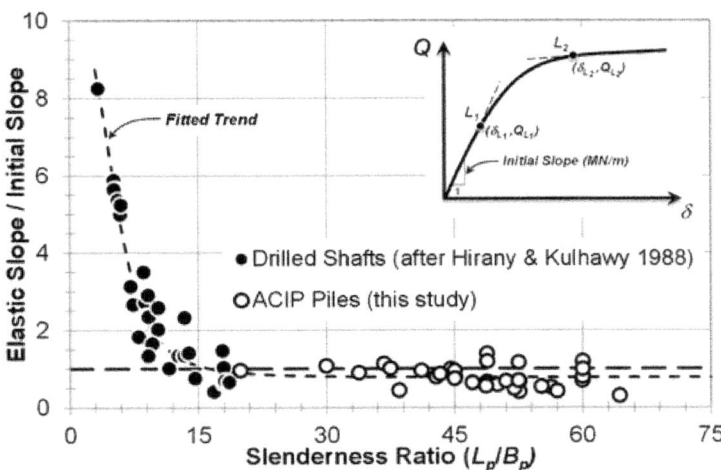

Figure 2: Variation of ratio of elastic compression line and initial slope re-

sulting from observed load–displacement curves with slenderness ratio for drilled foundations (inset: illustration of definitions associated with the L_1–L_2 **method**).

Chin (1970) proposed an extrapolation technique based on the general hyperbolic relationship described by Kondner (1963), giving rise to the Chin–Kondner extrapolation (CKE) method. The interpreted failure load derived from the CKE method is an asymptote of the Q–δ curve; therefore, $Q_{f,int}$ cannot actually be observed during a loading test (note that a plunging pile achieves and maintains, or loses, its maximum resistance, and therefore cannot be described by a hyperbola). In order to determine the interpreted failure load, the curve given by

$$Q = \frac{\delta}{k_1 + k_2 \delta} \qquad (3)$$

is fitted to the observed load–displacement data using ordinary least squares regression, where k_1 and k_2 are the fitting constant and coefficient respectively; $Q_{f,int}$ is then given by $1/k_2$. The CKE method is not currently specified by the IBC (2012).

The Hansen (1963) 90 Percent Criterion (90C) for interpreting failure loads sets $Q_{f,int}$ equal to the load corresponding to a displacement two times larger than the displacements associated with 90% of that load. It is noted that the IBC (2012) allows the use the 90C method, despite the extrapolation of the load–displacement curve commonly required for those instances where the pile head has been insufficiently displaced to achieve the displacement criterion (Fig. 1). The Chin–Kondner hyperbolic curve, the parabolic curve associated with the Hansen 80 Percent Criterion (Hansen, 1963), and the Decourt (2008) extrapolation method can prove useful in these instances. Another approach accepted by IBC (2012) is the Butler–Hoy Criterion (BHC; Butler and Hoy, 1977). This criterion sets $Q_{f,int}$ equal to the load given by the intersection of a line tangent to the initial slope of the Q–δ curve and a line tangent to the curve at a slope of 0.14 mm kN^{-1} (0.05 in. ton^{-1}) (Fig. 1). Similar to the 90C method, the extrapolation of the Q–δ curve is commonly required for the interpretation of static loading tests not carried to geotechnical failure.

For the example in Fig. 1, the four methods evaluated provide a range of $Q_{f,int}$ from about 1525 to 2630 kN (342 to 591 kips), spanning approximately 1100 kN (250 kips) or approximately 75% of the DOL interpreted failure load. It is worth noting that the example test pile exhibited a deflection hardening response for an additional 677 kN (152 kip) and 37 mm (1.4 in.) *beyond*the DOL capacity. In other words, the test pile can provide significantly more resistance than the failure load interpreted using the DOL method. Note that an interpreted failure load is typically factored by half or more to produce a working or allowable load, and this helps to keep displacements in check. For example, using a factor of safety (FS) of 2.0, the working load and displacement associated with the DOL method are 762 kN (171 kips) and 3.7 mm (0.15 in.) respectively, whereas a working load and displacement of 1315 kN (296 kips) and 9.7 mm (0.4 in.) results from applying a FS=2 to the CKE method. Because 40% of the pile head displacement at the CKE working load results from elastic compression (which would apply equally to all piles of similar length, loading, and stratigraphy for a given structure) assuming minimal toe resistance and typical granular load transfer profiles (i.e., triangular), the maximum differential settlement cannot be greater than approximately 6 mm (0.24 in.). This magnitude of differential settlement will produce angular distortions that may be considered tolerable for most structures (Zhang and Ng, 2007). Note that the preceding discussion is inappropriate for closely spaced or large pile groups overlying compressible layers. The remainder of this paper aims to generalize the preceding discussion and argue for and against various interpretation methods.

LOADING TEST DATABASE

In order to make satisfactory assessments of the selected failure load methodologies, the loading test database was limited to case histories with displacements greater than or equal to 5% of pile head diameter. The database compiled herein consists of 36 loading tests performed on ACIP piles installed in soil profiles consisting primarily of cohesionless soil layers: nine loading tests were compiled by Park *et al.* (2011), nine were compiled by McCarthy (2008) and one loading test was reported by O'Neill *et al.* (1999); the remaining 17 loading tests were contributed by member firms of the Deep Foundations Institute specifically for this study. The diameter, length and slenderness ratio L_p/B_p ranged from 350 to 460 mm (14–20 in.), 9.1 to 24.4 m (30–80 feet) and 20 to 64.3, with medians of 406 mm, 20.3 m and 48.7 respectively.

All of the load–displacement curves exhibited some amount of non-linearity, a requirement for admittance to the database. Maximum pile head displacements ranged from 18 to 50 mm, or 5 to 12.4% of pile head diameter, and the median displacement and normalized displacement, δ^*, was 33 mm (1.3 in.) and 8.2% respectively. In terms of the loading, the maximum applied test load ranged from 1288 to 3380 kN (290–760 kips), with a median maximum test load of 2666 kN (600 kips). Figure 3 presents histograms for the normalized maximum displacement and maximum applied test load to illustrate the spread of loading test data within the pile case history database.

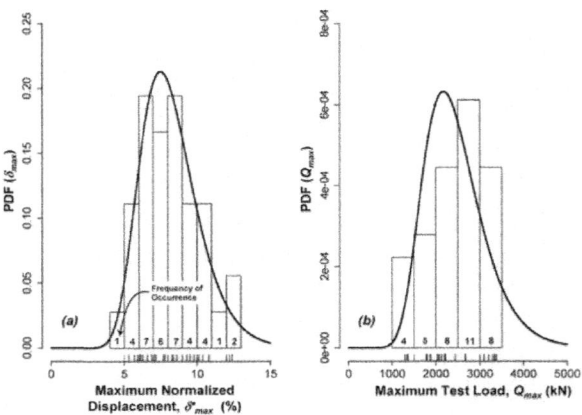

Figure 3: Histograms and fitted reference lognormal probability density functions (PDFs) of *a* **maximum normalized displacement and** *b* **maximum pile test load for pile case histories constituting load test database**.

ASSESSMENT OF INTERPRETED FAILURE CAPACITIES

The 36 loading tests were analyzed to produce interpreted failure loads associated with the DOL method, the MDOL method, the L_1–L_2 method, the CKE method, 90C method, BHC method and the STC method. Owing to the deviation of experimental data from a perceived ideal and occasional experimental noise, extrapolation techniques invariably require some degree of personal bias. For example, the first five Q–δ observations of the loading test data shown in Fig. 1 deviate significantly from the assumed hyperbolic response as shown in Fig. 4 where the same data are plotted with a transformed ordinate. Therefore, to estimate the CKE capacity, the hyperbolic curve was fit to all, but the first several Q–δ data pairs that significantly deviated from a linear trend in the transformed space. For those loading tests where there was insufficient displacement to determine the BHC and 90C capacities, a separate Chin-Kondner extrapolation of the load test data was carried out to give greater weight to the larger displacements and loads in an attempt to be most accurate over that portion of the Q–δ curve. Figure 4 illustrates the result this interpretative decision, where only the last 20 Q–δ observations were used to produce an extrapolation. Additionally, for those loading tests with an unload–reload, the CKE was used to bridge the gap in displacement between the first and second cycle to estimate an equivalent monotonic loading curve, as illustrated in Fig. 1. The elastic compression line for use with the DOL and MDOL method was computed using a composite modulus assuming a Young's modulus of 34.5 GPa for the concrete and 1% steel by area, which results in E_p=36.3 GPa. The following discussion addresses the comparison of interpreted failure loads and associated pile head displacements.

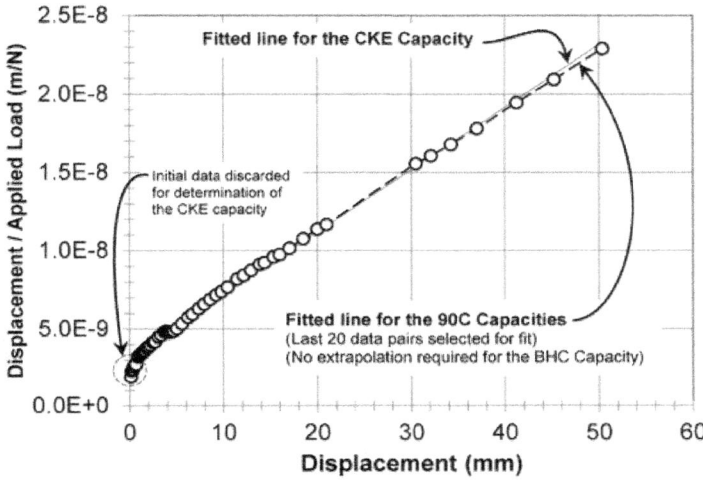

Figure 4: Illustration of fitting procedures for Chin–Kondner extrapolation (CKE) capacity and data weighting approach for limited extrapolation for case 32.

The interpreted failure capacity for each of the 36 pile case histories is presented in terms of the capacity (Fig. 5a) and in terms of the ratio of $Q_{f,int}$ to the DOL capacity (termed the DOL capacity ratio) as a reference and for ease of comparison (Fig. 5c and 5d). Table 1 presents the median, mean, range and coefficient of variation (COV, defined as the ratio of the standard deviation to the mean) for the DOL capacity and associated DOL capacity ratios. Figure 6 presents histograms illustrating the spread in the capacity ratios that are plotted in Fig. 5b and 5d. The DOL capacity ranged from 900 to 3230 kN (202–726 kips), with a median and mean capacity of 1752 and 1831 kN respectively. The COV in the DOL capacity, 32%, is in itself only an indicator of the spread of data in the database, and is not a performance measure. In comparison, the L_2 capacity, or the load at which the Q–δ curve begins to follow a secondary linear trend (Fig. 2), was approximately 7% larger than the DOL capacity on average, with a narrow dispersion representative of the proximity in load produced by these two interpretation methods. Likewise, the load multiplier for the STC method exhibited little dispersion, though it produced nearly the same estimate in capacity, on average, as the DOL method (Table 1). This is because there is very little difference between the elastic and initial stiffness for slenderness ratios greater than about 20 (Fig. 2).

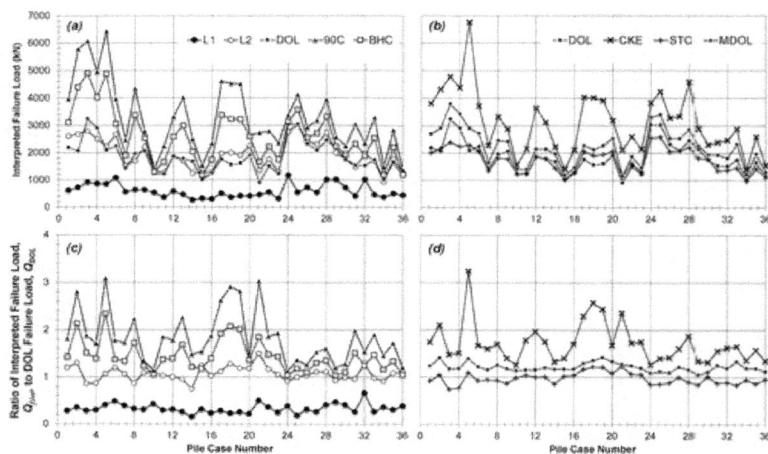

Figure 5: Result of capacity determinations: (a, b) Interpreted failure capacities, (c, d) ratio of interpreted failure capacity to the Davisson Offset Limit capacity.

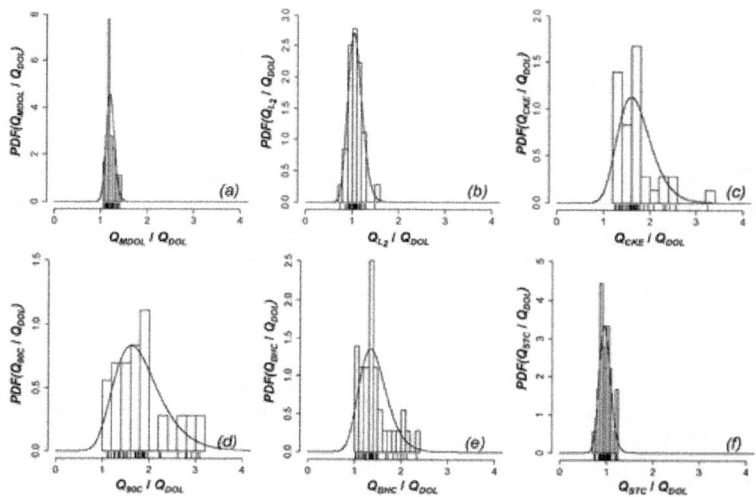

Figure 6: *a* **modified Davisson offset limit (MDOL) capacity;** *b* L_2 **capacity;** *c* **Chin–** Kondner extrapolation (CKE) capacity; *d* **90 Percent Criterion (90C) capacity;** *e* **Butler–Hoy Criterion (BHC)** capacity; *f* **slope tangent in compression (STC) capacity**.

Distributions of ratios of various interpreted failure load capacities to Davisson offset limit (DOL) method: fitted lognormal distribution is provided for reference.

Table 1: Comparison of summary statistics for selected interpreted failure capacity methods and associated displacements (where appropriate)

Performance measure	Median	Mean	Minimum	Maximum	COV/%
Davisson offset limit capacity Q_{DOL}/kN	1752	1831	900	3230	32·0
Ratio of L_1 capacity* Q_{L1}/Q_{DOL}	0·30	0·33	0·15	0·66	30·8
Ratio of L_2 capacity Q_{L2}/Q_{DOL}	1·07	1·07	0·74	1·50	14·1
Ratio of CKE capacity Q_{CKE}/Q_{DOL}	1·64	1·71	1·25	3·24	24·7
Ratio of 90C capacity Q_{90C}/Q_{DOL}	1·75	1·83	1·11	3·08	30·3
Ratio of BHC capacity, Q_{BHC}/Q_{DOL}	1·37	1·44	1·03	2·33	23·1
Ratio of STC capacity, Q_{STC}/Q_{DOL}	0·95	0·97	0·74	1·23	12·7
Ratio of MDOL capacity, Q_{MDOL}/Q_{DOL}	1·19	1·21	1·05	1·42	7·5
Davisson offset limit displacement δ_{int}/mm	14·7	14·8	9·5	22·6	21·6
Ratio of L_1 displacement δ_{L1}/δ_{int}	0·10	0·12	0·03	0·36	55·9
Ratio of L_2 displacement δ_{L2}/δ_{int}	1·14	1·16	0·44	2·20	31·5
Ratio of BHC displacement† $\delta_{BHC}/\delta_{int}$	2·03	1·95	0·91	3·65	30·0
Ratio of MDOL displacement‡ $\delta_{MDOL}/\delta_{int}$	1·54	1·57	1·27	1·99	8·9

*The L_1 capacity represents the departure of the Q–δ curve from the initial slope, and is therefore not a failure capacity.
†n=27 for calculation of δ_{BHC} due to insufficient displacement magnitudes for nine loading tests.
‡n=34 for calculation of δ_{MDOL} due to insufficient displacement magnitudes for two loading tests.

The MDOL, BHC, CKE and 90C produced interpreted failure loads that were 21, 44, 71 and 83% greater than the DOL capacity on average respectively. The COV in these load multipliers ranged from 7.5% for the MDOL method, which produced a moderate increase in capacity in comparison to the DOL capacity, to 30% for the 90C method, which often required extrapolation to estimate due to the larger displacements associated with this approach.

The average pile head displacement δ_{int} corresponding to the DOL capacity was 14.8 mm (0.6 in.), of which 5.3 mm (0.21 in.) constituted elastic compression of the pile element, on average, assuming typical load transfer profiles for granular soils. Pile head displacements associated with the CKE and 90C method could not be computed reliably due to the degree and hyperbolic nature of the extrapolation required. However, it was possible to compute the displacements associated with the L_2, MDOL and BHC capacities, which were equal to 1.16, 1.57 and 1.95 times greater than the DOL displacement on average respectively. In order to illustrate the scatter and global trend associated with various failure load interpretation methods, Fig. 7 presents the variation of the DOL capacity normalized interpreted failure load against the DOL intersection displacement normalized displacements $\delta_{f,int}$ associated with $Q_{f,int}$ to illustrate the relationship between the capacity ratio and the displacement ratio for the pile case histories in this study.

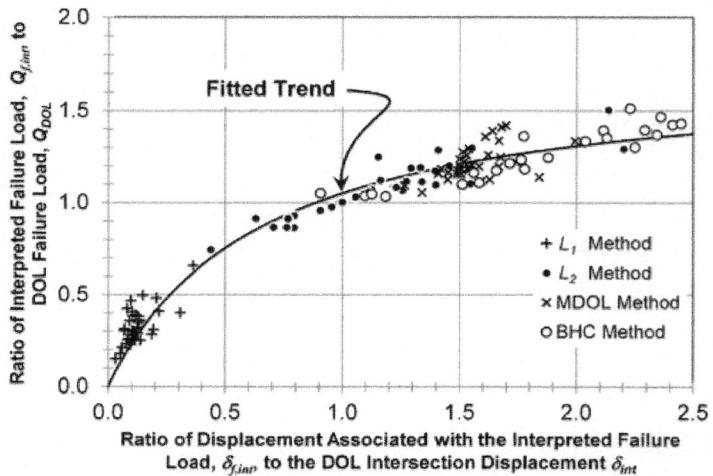

Figure 7: Variation of Davisson offset limit capacity normalized interpreted failure loads $Q_{f,int}$ **with intersection displacement normalized displacements associated with various failure load interpretation methods** $\delta_{f,int}$: fitted hyperbola is provided for reference.

It is emphasized that it is desirable to achieve large displacements during a static loading test such that critical soil responses can be observed (e.g. deflection softening or hardening) and an accurate estimate or approximation of the "true ultimate resistance" can be obtained. Additionally, the reliability of many failure load interpretation criteria reduces when significant extrapolation is required. Thus practitioners are encouraged to displace test piles to as large a magnitude as possible, as few building officials will accept significant extrapolation of the Q–δ curve and often only with considerable justification. However, the selection of a failure load interpretation method should not needlessly handicap the efficiency of the foundation. Consider the pile head displacements associated with the BHC: the average displacement at the BHC capacity was 96% larger than the displacement at the DOL capacity (Table 1), or 28.3 mm (1.1 in.), larger than typically acceptable for structure foundations. However, this magnitude of displacement is associated with the estimate of the failure load, not the working load. Assuming a FS=2, the average displacement pile head displacement reduces to 7.8 mm (0.31 in.); this observation can be confirmed by considering the non-linearity in Figs. 1 and 7. Furthermore, when neglecting the contribution of elastic compression associated with the loading and assuming the typical load transfer profile for granular soils, the mean differential pile head displacement reduces to 4.2 mm (0.16 in.). Few structures require maintaining differential settlements below this stringent magnitude. Thus, general arguments for excessively conservative failure load interpretation methods based on pile head displacement appear unwarranted.

DISCUSSION AND RECOMMENDATIONS

Analysis of the ACIP pile database has provided a quantitative basis for the comparison of several IBC accepted and other failure load interpretation methods. Qualitatively, an appropriate failure load interpretation method should (Hirany and Kulhawy, 1988): be independent of subjective judgment (e.g. a mathematical rule) and plotting scale (Chellis 1961; Fellenius,1980), consider differential settlement of the pile foundation only and neglect elastic compression (Fuller and Hoy, 1970), consider the diameter of the foundation as a proxy for scaling effects (Vesic, 1977), consider pile length and stiffness (Fellenius, 1980), and not use extrapolation unless the Q–δ curve indicates the approach of an asymptote (Leonards and Lovell, 1979). These factors, among others, are considered in the following discussion.

Methods accepted by IBC

Despite the use of the DOL method as a convenient approach for presenting the capacity comparisons described herein, it does not represent an appropriate approach for estimating the interpreted failure load. The DOL method assumes that the pile behaves as a toe bearing element only, such that is can be modeled as a free standing fixed based column (Davisson, 1972; NeSmith and Siegel, 2009). Drilled foundations provide significant shaft resistance, with the large majority of the total resistance provided by the shaft at working loads. ACIP piles are not driven using impact methods; therefore, the use of a wave equation based model is inappropriate. As described earlier, the stress history of a soil element below a drilled foundation is considerably different than that of driven pile: an element of soil below the toe experiences an increase in stress relative to its initial stress state, even if relaxation, due to a dilative soil response, occurs, whereas a soil element below a drilled foundation experiences unloading and reloading to its more or less initial stress state. The soil below a drilled foundation is not prestressed unless postgrouting is performed. Additionally, the DOL method requires the calculation of the elastic compression of the foundation, and this is not a straightforward task for drilled foundations due to the strain dependent constitutive nature of the steel reinforced concrete composite. Strain gages above or near the ground surface are required to accurately backcalculate the appropriate modulus for use with computing the elastic shortening of a drilled foundation (Fellenius, 1989, 2012; Stuedlein and Gurtowski, 2012), as the typical ACI regression equation for elastic modulus (ACI 318; ACI, 2011) can provide errors on the order of 25%

or more depending on the water/cement ratio, admixtures, and aggregate size and mineralogy. Because the modulus of the concrete varies with the imposed strain and along the length of the shaft due to load transfer, it is incorrect to select a single value to represent the compressibility of the pile, in addition to the error introduced through the use of elastic modulus correlations. Thus, the soil and pile models envisioned by Davisson (1972) and developed for use with driven piles are inappropriate for ACIP piles and other drilled foundations.

The other two methods accepted by the IBC include the 90C and the BHC. The 90C method is based on a mathematical rule, but makes no assumptions regarding the stiffness or diameter of the pile. The 90C method may be considered relatively efficient for ACIP piles as it yielded an 83% increase in interpreted failure load on average as compared to the DOL method (Table 1), with a relative increase that ranged from 1.11 to 3.08 times greater than the DOL capacity. However, significant extrapolation of the Q–δ curve was commonly required to estimate the 90C capacity in this investigation, regardless of the magnitude of the maximum displacement that occurred during the loading test. Extrapolation, though easily performed, must be used carefully and judiciously, and only if sufficient knowledge and certainty in the stratigraphy exists. For example, the extrapolation of the capacity of a drilled foundation bearing on glacially overridden materials seems reasonable, as all other soil units below the bearing stratum must also be necessarily overridden. However, the hydrogeological environment must also be considered, as a pressurized aquifer could readily provide smaller toe bearing resistances than that estimated assuming a hydrostatic pore pressure distribution. Extrapolation of a Q–δ curve to an ultimate geotechnical resistance for a friction pile, or toe bearing pile tipped into a thin dense layer overlying a compressible deposit, must be avoided. Because the displacements associated with the 90C capacity could not reliably determined for the database, it is difficult to make an informed and general argument for the reliable use of this method in practice.

On the other hand, the BHC method provided a systematic approach that was largely independent of user interpretation and an interpreted failure load that was significantly less overconservative than the DOL method. The average increase in capacity was 44% over the DOL method, with a relative increase ranging from 1.03 to 2.33 times greater than the DOL capacity. For the 27 pile loading tests that were displaced sufficiently to characterize $\delta_{f,int}$ for the BHC method, the average $\delta_{f,int}$ was 95% greater than the DOL method, indicating that significantly more toe resistance was mobilized, and therefore, more representative of a true ultimate resistance (i.e. geotechnical failure). This is considered desirable, as it allows the pile to be used more efficiently. Recall the earlier discussion of displacements at the working load and associated differential settlements: even if a structure is supported on foundations of different lengths, diameters, or in different soil deposits, the contribution of the displacement equal to the minimum elastic compression can be neglected in the response of the structure. Thus, failure load interpretation methods that produce larger factored resistances do not necessarily lead to increased risk of poor structural performance. Considering the desirable elements in a failure load interpretation method, the BHC method is independent of judgment (i.e. specifies a quantitative limiting settlement rate to prevent excessive creep), considers pile stiffness (through use of the initial slope of the Q–δ curve), and did not require significant extrapolation for the piles evaluated herein. Owing to its ease of use, systematic implementation free from interpretational bias, and the resulting reasonable and cost effective capacities, the BHC is the most satisfactory failure load interpretation method for use with ACIP piles that is accepted by the IBC.

Methods not Explicitly accepted by IBC

The IBC (2012) allows the use of "Other methods approved by the building official;" however, this prerogative is not frequently invoked in the experience of the authors. Nonetheless, improved efficiencies could be realized by using the L_1–L_2 and modified MDOL, which typically provided greater interpreted failure capacities than the DOL method. Comparison of these two methods indicates that the MDOL capacity was always larger than the DOL method (ranging from 5 to 42%), and larger than the L_2 capacity on average. One benefit of the MDOL method is that the diameter of the drilled foundation is directly incorporated into the failure load interpretation procedure. Unfortunately, the MDOL requires the use of an elastic compression line and use of an assumed elastic compression line is not recommended for reasons described above. An advantage of the L_2 capacity is that it requires no assumptions regarding the elastic stiffness of the foundation element. Additionally, Hirany and Kulhawy (1988) point out that the L_2 capacity correlates with the mobilization of a certain magnitude of toe resistance. However, the L_1–L_2 method requires sufficient displacement to accurately determine the L_2 capacity. Indeed, the authors found that significant judgment was required to determine the L_2 capacity, and this was considered the main disadvantage of this method. With regard to the magnitude of displacements, the MDOL procedure generally produced larger displacements than those associated with the L_2 capacity, and this indicates that it represents a better estimate of the "true ultimate resistance" of the pile foundation.

Recommended Approach for Interpretation of Failure Load for ACIP Piles

In the framework of deterministic pile design, the appropriate pile capacity value inferred from a failure load interpretation method is the maximum capacity possible that, when factored appropriately, maintains serviceability to the structure in consideration of differential settlement. Additionally, an appropriate interpretation method should be free from the effects of judgment or other sources of subjective influence. A review of the various factors controlling the interpretation of failure loads from static loading tests set in the context of the ACIP pile database indicated that the commonly used Davisson (1972) offset limit method is inappropriate for drilled foundations. The MDOL (Kyfor et al., 1992) presents an advantageous alternative; this method would be recommended if the initial slope tangent of the load–displacement curve was substituted for the elastic compression line, eliminating the error associated with estimating an elastic modulus. Assuming that displacements do not exceed the required serviceability limit, and based on the requirement to use a sound, objective evaluation procedure that maximizes cost effectiveness, the BHC is recommended for the interpretation of the failure load for static loading tests on ACIP piles.

CONCLUSIONS

This paper describes an effort to evaluate the appropriateness of commonly used failure load interpretation methods for ACIP piles. Methods currently accepted by the IBC (2012) were emphasized, but other methods with historical and contextual precedent were also considered. A database of 36 ACIP pile loading tests with displacements equal to or greater than 5% of the pile head diameter was assembled and analyzed. Because of its unnecessarily conservative nature and corresponding increase in cost associated with a larger number of piles, larger pile diameters, and/or pile caps, and critically, the deviation of its assumptions from widely accepted engineering principles, this study concluded that the original DOL method is inappropriate for ACIP and other drilled foundations. The use of the Davisson Offset Limit should be discontinued in practice for ACIP and other drilled foundations. Although other methods presented various advantages, the BHC was found to represent the most appropriate method from the perspectives of engineering soundness and pile efficiency. Owing to its current acceptance in the IBC, this method is recommended for ACIP piles and other drilled foundations.

In closing, the interpretation of the failure load from a static pile loading test is an important task among several required for successful design of piled foundations. As usual, geological details are critical; therefore, significant effort must be made to identify disadvantageous geological conditions, such as deep compressible soil layers or non-hydrostatic conditions. The preferred displacement based design must consider these and other mitigating factors as part of a comprehensive design program.

REFERENCES

1. ACI. 2011. Building code requirements for structural concrete and commentary, ACI Committee 381, American Concrete Institute, Farmington Hills, MI, USA.

2. ASTM Standard D 1143/1143M. 2013. Standard test methods for deep foundations under static axial compressive load, ASTM International, West Conshohocken, PA, USA.

3. Brown DA, Turner JP and Castelli RJ. 2010. Drilled Shafts: Construction Procedures and LRFD Design Methods, Geotechnical Engineering Circular No. 10, Report No. FHWA NHI-10-016, Federal Highway Administration, 970 pp..

4. Butler HD and Hoy HE. 1977. The Texas quick-load method for foundation load testing, User's manual, Report no. FHWA-IP-77-8, Federal Highway Administration, Washington, DC, USA, 1977.

5. Chellis RD. 1961. Pile foundations, 2nd edn; New York, McGraw Hill Book Company.

6. Chin FK. 1970. Estimation of the ultimate load of piles not carried to failure, Proc. 2nd Southeast Asian Conf. on 'Soil engineering', Singapore, June, Southeast Asian Society of Soil Engineering, 81–90.

7. Davisson MT. 1972. High capacity piles, Proc. Lecture series on innovations in foundation construction, 81–112; Chicago, IL, Soil Mech. Div., Illinois Sec., ASCE.

8. Davisson MT. 1993. Negative skin friction in piles and design decisions, Proc. 3rd Int. Conf.on 'Case histories in geotechnical engineering', St Louis, MO, USA, June, University of Missouri-Rolla, 1793–1801.

9. Decourt L. 2008. Loading tests: interpretation and prediction of their results, Proc. Symp. Honoring Dr. John H. Schmertmann for His Contributions to Civil Engineering at Research to Practice in Geotechnical Engineering Cong. 2008, New Orleans, LA, USA, March, ASCE, 452–470.

10. Fellenius BH. 1975. Test loading of piles and new proof testing procedure, *J. Geotech. Engrg. Div., ASCE*, 101, (9), 16–30.

11. Fellenius BH. 1980. The analysis of results from routine pile load tests, *Ground Eng.*, 13, (6) 19–31.

12. Fellenius BH. 1989. Tangent modulus of piles determined from strain data, Proc. The 1989 Foundation Cong. of the Geotechnical Engineering Division, (ed. Kulhawy F H), Vol. 1, 500–510; New York, ASCE.

13. Fellenius BH. 2012. The red book: the basics of foundation design, Electronic Edition, http://Fellenius.net, (accessed 15 October 2013).

14. Fuller FM and Hoy HE. 1970. Pile load tests including quick load test method, conventional methods, and interpretation, Research Record 333, Highway Research Board, Washington, DC, USA, 74–86.

15. Hansen JB. 1963. Discussion of "hyperbolic stress–strain response: cohesive soils", *J. Soil Mech. Found. Div., ASCE*, 89, (SM4), 241–242.

16. Hirany A and Kulhawy FH. 1988. Conduct and interpretation of load tests on drilled shaft foundations: detailed guidelines, Report EL5915, Electric Power Research Institute, Palo Alto, CA, USA.

17. Hirany A and Kulhawy FH. 1989. Interpretation of load tests on drilled shafts – 1: axial compression, in Foundation engineering: current principles & practices, GSP 22, (ed. Kulhawy F H), 1132–1149; New York, ASCE.

18. IBC. 2012. International Building Code, International Code Council, Falls Church, VA, USA.

19. Kondner RL. 1963. "Hyperbolic Stress-strain Response of Cohesive Soils," Journal of the Soil Mechanics and Foundations Division, ASCE, Vol. 89, SM1, pp 115–143.

20. Kulhawy FH. 2004. On the axial behavior of drilled foundations. *GeoSupport: Drilled Shafts, Micropiling, Deep Mixing, Remedial Methods, and Specialty Foundation Systems, GSP 124*, J.P. Turner, and P.W. Mayne, eds., ASCE, Reston, VA, 1–18.

21. Kulhawy FH and Hirany A. 2009. Interpreted failure load for drilled shafts via Davisson and L_1–L_2, in Contemporary topics in deep foundations, GSP No. 185, 127–134; Reston, VA, ASCE.

22. Kyfor ZG, Schnore AR, Carlo TA and Baily PF. 1992. Static testing of deep foundations, Final Report FHWA-SA-91-042, Federal Highway Administration, Washington, DC, USA, 174.

23. Leonards GA and Lovell D. 1979. Interpretation of load tests on high capacity driven piles, in Behavior of deep foundations, Special Technical Publication 670, 388–415; Philadelphia, PA, ASTM.

24. McCarthy DJ. 2008. Empirical relationships between load test data

and predicted compression capacity of augered cast-in-place piles in predominately cohesionless soils, MSc thesis, Univ. of Cent. Florida, Orlando, FL, USA.

25. NeSmith WM and Siegel TC. 2009. Shortcomings of the Davisson offset limit applied to axial compressive load tests on cast-in-place piles, in Contemporary topics in deep foundations, GSP No. 185, 568–574; Reston, VA, ASCE.

26. O'Neill MW and Reese LC. 1999. "Drilled Shafts: Construction Procedures and Design Methods," Federal Highways Administration, Washington, D.C., 758 pp.

27. O'Neill MW, Vipulanandan C, Ata A and Tan F. 1999. Axial performance of continuous-flight-auger piles for bearing. *Texas Dept. of Transportation report no. 7-3940-2*, Center for Innovative Grouting Materials and Technology: University of Houston.

28. Park S, Roberts LA and Misra A. 2011. Static load test interpretation using the t-z model and LRFD resistance factors for auger cast-in-place (ACIP) and drilled displacement (DD) piles, *Int. J. Geotech. Eng.*, 5, (3), 283–295.

29. Stuedlein AW and Gurtowski TM. 2012. Reliability of shaft resistance of augered cast-in-place piles in granular soils, in Full-scale testing and foundation design: honoring Bengt H. Fellenius, GSP No. 227, 722–736; Reston, VA, ASCE.

30. Stuedlein AW, Neely WJ and Gurtowski TG. 2012. Reliability-based design of augered cast-in-place piles in granular soils, *J. Geotech. Geoenviron. Eng., ASCE*, 138, (6), 709–717.

31. van Weele AF. 1982. Which load is allowable on a given foundation pile when its actual load-settlement behavior is available through a load test?, Amici et Alumni. Em Prof. Dr. Ir E. E. De Beer, 287–296.

32. Vesic AS. 1977. Design of pile foundations, Synthesis of Highway Practice 42, Transportation Research Board, Washington, DC, USA.

33. Zhang L and Ng A. 2007. Limiting tolerable settlement and angular distortion for building foundations, in Probabilistic applications in geotechnical engineering, GSP No. 170, 1–11; Reston, VA, ASCE.

Chapter 11

FIBER BRAGG GRATING-BASED PERFORMANCE MONITORING OF PILES FIBER IN A GEOTECHNICAL CENTRIFUGAL MODEL TEST

Xiaolin Weng[1], Jianxun Chen[2], and Jun Wang[3]

[1]Key Laboratory for Special Area Highway Engineering of Ministry of Education, Chang'an University, Xi'an, Shaanxi, China

[2]School of Highway, Chang'an University, Xi'an 710064, China

[3]China Railway First Survey and Design Institute Group Ltd., Xi'an 710064, China

ABSTRACT

In centrifugal tests, conventional sensors can hardly capture the performance of reinforcement in small-scale models. However, recent advances in fiber optic sensing technologies enable the accurate and reliable monitoring of strain and temperature in laboratory geotechnical tests. This paper outlines a centrifugal model test, performed using a 60 g ton geocentrifuge, to investigate the performance of pipe piles used to reinforce the loess foundation below a widened embankment. Prior to the test, quasidistributed fiber Bragg grating (FBG) strain sensors were attached to the surface of the pipe piles to measure the lateral friction resistance in real time. Via the centrifuge actuator, the driving of pipe piles was simulated. During testing, the variations of skin friction distribution along the pipe piles were measured automatically using an optical fiber interrogator. This paper represents the presentation and detailed analysis of monitoring results. Herein, we verify the reliability of the fiber optic sensors in monitoring the model piles without affecting the integrity of the centrifugal model. This paper, furthermore, shows that lateral friction resistance developed in stages with the pipe piles being pressed in and that this sometimes may become negative.

INTRODUCTION

Based on such advantages as sound quality, bearing capacity, and low construction noise, the application of the prestressed concrete pipe pile has extended from the coastal, soft-soil regions to the inland, collapsible loess areas. Researchers, however, have recently focused more on the lateral friction resistance of pipe pile than on construction and inspection, which were of interest in early years. Shi [1] conducted a test by placing strain gauges in both ends of the pile and at the interfaces between different soil layers to measure both lateral friction resistance and bearing capacity. However, the thickness of the pipe influences this method to the extent of causing appreciable error due to the loss in friction resistance. Leng et al. [2] proposed embedding steel plates affixed to strain gauges during the prefabrication of pipe piles. Xiaokui [3] attempted to use Brillouin Optical Time Domain Reflectometry (BOTDR) to perform a load test on prestress high concrete (PHC) pipe piles that had already been hammered into the ground, but he found it impractical because of the excessively long sampling time. Again by use of BOTDR, Xing et al. [4] conducted a static load test on PHC pipe piles and succeeded in monitoring the load variation at each level. When Xu [5] performed strain monitoring, by placing FBG sensors on the surface of the static pressure piles, he succeeded in measuring end resistance and lateral friction resistance separately.

The advantages of optical fiber sensors over conventional monitoring techniques include immunity to electromagnetic interference, small size and lightweight construction, and access to different measurands, such as strain, temperature, vibration, and specified chemicals [6]. The optical fiber sensors can also be multiplexed, meaning that more than one sensor can be integrated along a single optical fiber. This multiplexing capability can enable a distributed mapping of the structure to be monitored [7].

In the past decade, a series of fiber optic sensing technologies have been developed and applied to health monitoring of various geotechnical related structures, such as foundations, tunnels [8], caverns [9], and slopes [10]. Weng and Wang [11] reported the monitoring results of a physical model test of pavement instrumented by horizontally embedded FBG strain sensors. The response of the pavement due to differential settlement of embankment was presented and analyzed. Recently, Zhou et al. [12] utilized the same technology to perform in situ three-dimensional health monitoring of a pavement structure. In this testing process, researchers from Chang'an University used a 60 gt geocentrifuge and a subgrade model to study the performance of pipe piles in the collapsible loess foundation when placed under the load of widened subgrade. The centrifuge actuator was employed to simulate the pressing-in of the static pressure pile in the centrifugal field. Our researchers monitored the strain of piles using FBG sensors during different time periods—when piles were pressed in, subgrade put in service, or collapsibility generated within the loess foundation. The result of this study can provide reference for the design and construction of pipe piles in loess areas.

BRAGG GRATING SENSORS

The sensing functioning of FBG was first discovered upon the formation of photogenerated gratings in germanosilicate optical fiber by Hill et al. [13]. Table 1 lists a feature comparison between the conventional and FBG sensors. The Bragg grating is written into a segment of germanium-doped (Ge-doped) single-mode fiber in which exposure to a spatial pattern of ultraviolet (UV) light forms a periodic modulation of the core refractive index. Figure 1 illustrates the working principle of an FBG sensor. According to Bragg's law, when a broadband source of light has been injected into the fiber, FBG reflects a narrow spectral part of light at a certain wavelength [14]:

$$\lambda_B = 2n_{\text{eff}}\Lambda,$$

(1)

where λ_B is the Bragg wavelength, typically 1510 to 1590 nm (1 nm = 10^{-9} m); n_{eff} is the effective core index of refraction; and Λ is the period of the index modulation.

Table 1: Physical properties of the test soil

Void rate	Saturation %	Dry density g/cm³	Wet density g/cm³	Natural water content %	Coefficient of collapsibility	Coefficient of Nonuniformity Cu	Coefficient of curvature (Cc)
18.62	1.85	1.39	1.57	12.84	0.086	8.02	1.35

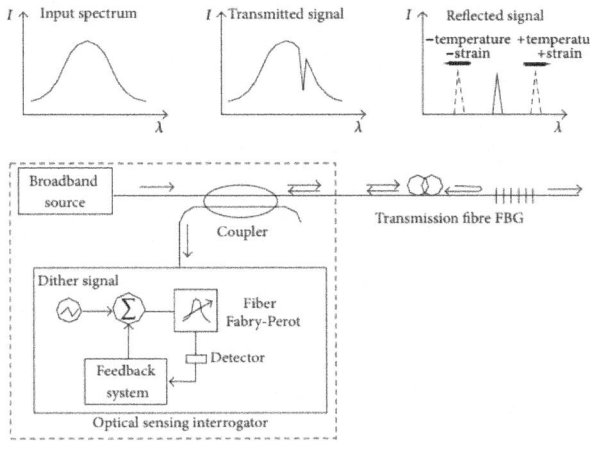

Figure 1: Strain and temperature sensing of an FBG sensor.

Through physical or thermal elongation of the sensor segment, as well as through change in the fiber refractive index caused by photoelastic and thermo-optic effect, the Bragg wavelength will change linearly with strain and temperature. Considering a standard single-mode silica fiber, λ_B changes linearly with the applied strain Δe and temperature ΔT. This relationship is given by Kersey et al. [15] as

$$\frac{\Delta \lambda_B}{\lambda_{B0}} = c_\varepsilon \varepsilon + c_T \Delta T,$$

(2)

where λ_B is the original Bragg wavelength under strain free and 0°C condition, $\Delta \lambda_B$ is the variation in Bragg wavelength due to the applied strain and temperature, and c_ε and c_T are the calibration coefficients of strain and temperature. The typical strain and temperature accuracy of a bare FBG sensor are 1 µε and 0.1°C, respectively.

Because the FBG sensor is sensitive to both strain and temperature, separating the effect of temperature from the strain monitoring data becomes a key problem in data analysis. This temperature-compensation problem can be achieved by adding an FBG sensor or a conventional temperature sensor to the same temperature field. Once the temperature is measured, the mechanical strain can be corrected as follows [9]:

The FBG package is one of the key points in strain monitoring. Only 250 µm cross, the bare fiber is very thin and exerts little influence on the monitored structure. In this test, fibers were adhered directly to the surface of the seamless steel pipe using ALTECO epoxy resin. As shown in Figure 2, twelve FBG sensors are symmetrically arranged from the point 15 mm below the top in two lines along the steel pipe at an interval of 20 mm. Readings from the pair of sensors at the same height will be averaged (the six pairs are numbered FBG1–6 from the top down). Figure 3 shows a pipe pile model with sensors already packaged. The sensors are affixed in parallel with the steel pipe so that they develop the same strain. A dynamic optical demodulator, SM130 of Micron Optics, is applied to measure the variation of wavelength of the FBG. The SM130 is a high-speed, heavy-duty measuring system designed primarily for use in mechanical testing. Capable of handling multiple sensors, this system applies to the measurement of strain, temperature, pressure, acceleration, and more. The SM130 has a top scan frequency of 1 kHz, which well meets the requirement of strain measurement at various pressing rate of piles.

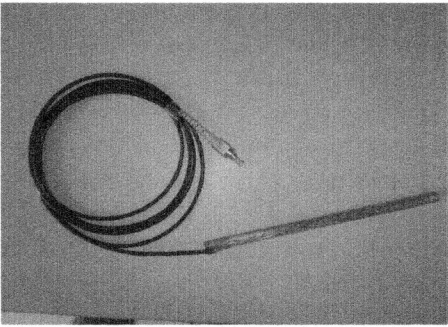

Figure 2: Model pipe pile with sensors packaged.

Figure 3: Model pipe pile with sensors packaged.

CENTRIFUGAL EXPERIMENTS

The soil for this test was taken from the suburb of Xianyang, which has an optimum water content of 14.0% and a maximum dry density of 1.83 g/cm³ as the result of compaction testing. According to the consolidated undrained test, conducted under different levels of compression, the cohesion and the internal friction angle of the sample soil are, respectively, obtained as 51 kPa and 22°. Table 1 shows the other physical properties of the test soil.

The model test is designed according to the centrifuge actuator. The model is prepared in box sizes 740 mm × 560 mm × 460 mm on a scale of 1/60, as shown in Figure 3. The foundation contains a bottom layer Q3 of loess and a top layer Q1 of original soil, both placed and compacted in layers. The compaction intensity for Q3 is adjusted to obtain a desired elastic modulus. The original soil for Q1 is used in order to maintain collapsibility of the foundation. Dry silver sand is placed between the soil and the model box to reduce the error caused by cracking. Both the new and old 110 mm high subgrades are placed and compacted in as many layers as possible to minimize foundation damage. At the beginning of the test, the placement of the old subgrade is simulated. Following this, the centrifuge is run at its top acceleration, 60 g, for a long period to ensure the old foundation soil material consolidation fully. After that, researchers stop the centrifuge, cut steps on the slope of the old subgrade, as shown in Figure 3, and then start the foundation treatment by using pipe piles. These seamless steel pipe piles—170 mm in length, 8 mm in diameter, and 1 mm in thickness—are arrayed at an interval of 30 mm. They represent prototypes that are 10 m in length, 480 mm in diameter, and 60 mm in thickness and are arrayed at an interval of 1.8 m, as shown in Figure 4. Coarse sand is chosen for the bedding cushion. After the foundation treatment is complete, the new subgrade is placed in layers after the completion of foundation treatment. Next, the centrifuge is run in six steps, to its top acceleration of 60 g, with runtimes of, respectively, 43.2, 10.8, 4.8, 2.7, 1.7, and 1.2 minutes. This simulates new subgrade placed in six layers with duration of three days for each. The centrifuge is run at 60 g for 438 minutes, which is equivalent to three years of service of the new subgrade; the strain of piles is monitored during this period. Later, the new subgrade is humidified to cause collapsible deformation. Strain of the piles is again monitored.

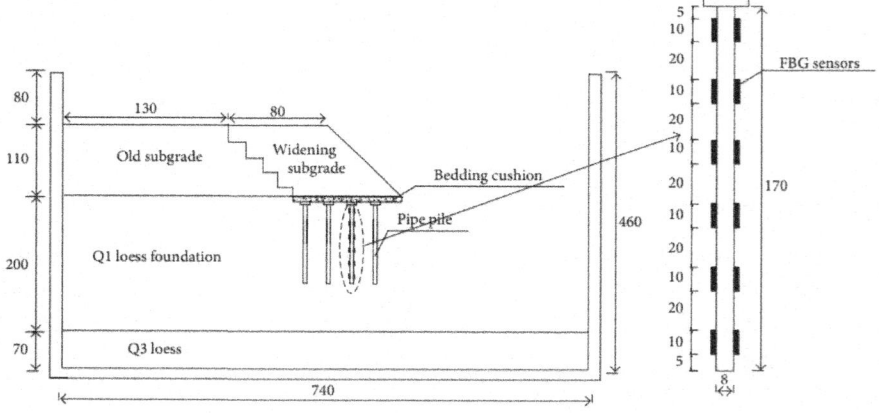

Figure 4: Illustration of the model test (unit: mm).

Design of the Pipe Pile

The pipe pile for the model test, in accordance with the similarity theory, is made of the same material (i.e., concrete and reinforcing steel) used in the real one. However, the similarity theory also determines the diameter of the model pipe pile to be 10 mm (600/60), which is almost impossible to obtain by casting concrete; therefore, a substitute material is needed. Hence the 10 mm seamless steel pipe is applied in this test to simulate the real pipe pile, which is 10 m in length, 60 cm in diameter, and 30 cm in thickness with an elastic modulus of 60 GPa. Figure 5 shows the result of the unconfined compression test of the steel pipe. By fitting the stress-strain curve, an elastic modulus of 55 GPa is obtained for the model pile, which achieves a similarity ratio of 1 to that of the real one. This has effectively reflected the feasibility of replacing the concrete pipe with seamless steel pipe.

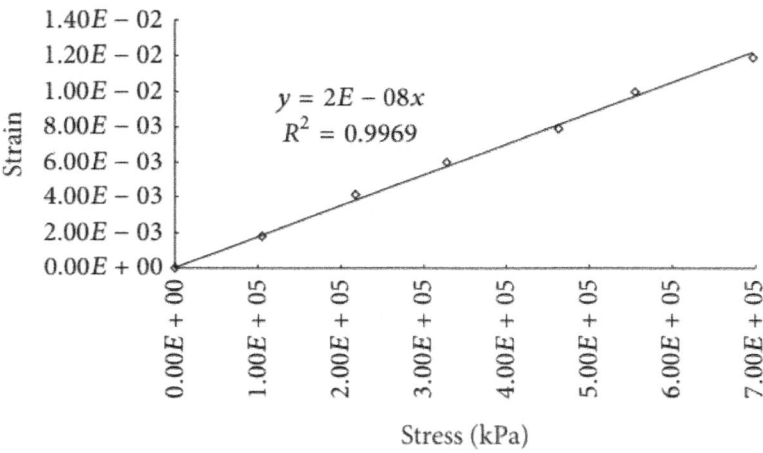

Figure 5: Result of unconfined compression test of seamless steel pipe.

Pressing-In of the Pipe Piles

Simulating the construction of jacked pipe piles, the pipe pile is pressed in within the centrifugal field by the 60 g ton geocentrifuge actuator of Chang'an University, as shown in Figure 6. The actuator is composed of the mechanical system and the electric system; the former includes the robotic arm and model box while the latter includes the computer and monitoring systems along with the power control cabinet. The actuator has four axes (i.e., X, Y, Z, and C), each independently driven by a motor to achieve four-axis positioning. As Figure 7 shows, it also includes a tool designed to press piles. It takes nine times to press the whole 170 mm long model pile into soil. With the exception of the final time, the pile goes 20 mm deeper each time; it has only 10 mm to go the last time. The model piles are arranged at different ratio S/d (S is the distance between piles and d is the pile diameter) of 1, 2, and 3 to evaluate the friction resistance of collapsible loess foundation against various pile arrangements. Figure 8 shows model piles being pressed in by the actuator, and Figure 6 shows the model piles already driven in place at an S/d ratio of 4.

Figure 6: Centrifuge actuator.

Figure 7: Pipe piles press-in process.

Figure 8: Finish of jacked pipe piles (S/d=4).

Water Log Infiltration System

A plexiglass box with two chambers was used in the experimental test setup to simulate water spill into soil from an underground storage tank, as shown in Figure 9. This type of experimental test setup was first used by Esposito and Allersma [16] and Pasha et al. [17]. The setup included a tank and a supply of the flushing liquid, along with a dispenser box to inject the flushing liquid into loess soil. The tank was fixed onto the centrifuge platform, and the dispenser was placed on top of soil in the strongbox. The chamber was connected to the air-pressure supply. At the appropriate time, when the centrifuge was spinning, air pressure was applied to the tank to push the required volume of water into the dispenser. The dispenser was refilled many times by this process until the total required flushing volume had been injected.

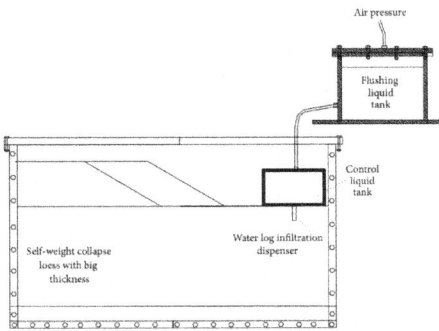

Figure 9: Schematic of centrifuge experiments setup for water log infiltration.

RESULTS AND DISCUSSION

The lateral friction resistance of the pipe piles is shown in Figure 10. This reveals that the pipe pile behaves in the exact way of a friction pile in the loess foundation. Before subgrade filling, the pile bears only its dead load and the lateral friction resistance stays at a low level, with a maximum value of approximately 20 kPa. Shortly after the subgrade filling, the pipe pile bears the most substantial part of the subgrade load and moves down, relative to the surrounding soil. The lateral friction resistance increases substantially, up to 80 kPa near the top, and decreases along the way, down to a minimum of 40 kPa. When one year has passed since subgrade filling, the lateral friction resistance begins to show a sharp decrease, caused by the compression of the soil body between the piles; this compression, in turn, results from its participation in bearing subgrade load. The pipe pile exhibits a negative correlation between the lateral friction resistance and the ratio of S/d. The smaller the ratio, the greater the friction resistance, as the soil between the piles is more intensively compressed. The study results are basically similar with Kong et al. [18] and Huang et al. [19].

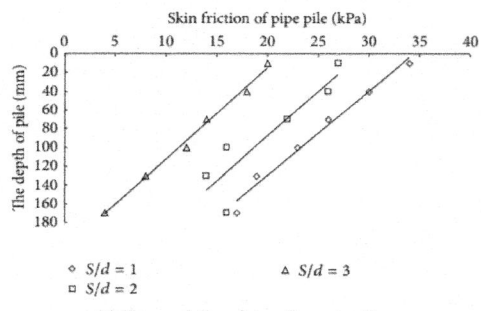

(a) The completion of pipe pile construction

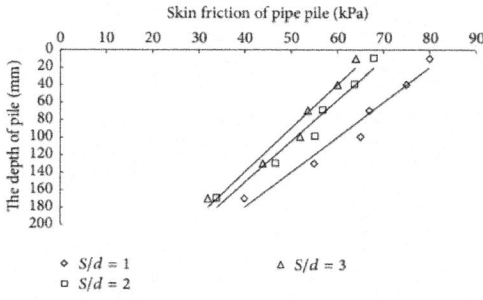

(b) The completion of subgrade filling

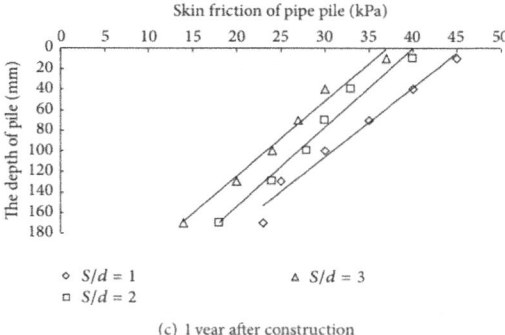

(c) 1 year after construction

Figure 10: Skin friction distributions along the pile depth.

Figure 11 shows the lateral friction resistance of the pipe pile in wetted, collapsible loess foundation. Although the pipe pile treatment neutralizes the collapsibility of the original loess foundation to a certain degree, collapse still occurs in various degrees—depending on the intensity and duration of water soaking on the surface—and causes negative friction. The longer the soaking time, the greater the negative friction. The negative-friction resistance decreases along the way down, eventually reaching a neutral point. According to Figure 11, after three days of water soaking, negative friction develops within the area 20 mm below the top of pile. As soaking time increases, the negative friction also increases, and the neutral point moves gradually down and finally stops at a certain depth. For example, after ninety days of water soaking, the ratio between the depth of neutral point and the length of pipe pile is $0.35 \sim 0.41$. According to Figure 11(a), the negative friction increases only slightly after ninety days of soaking beyond its friction after three days of soaking. In comparison, this negative friction in Figure 11(c) goes up to 45 kPa, which indicates that dense arrangement of piles can improve the intensity of compression, greatly reducing the collapsibility of soil.

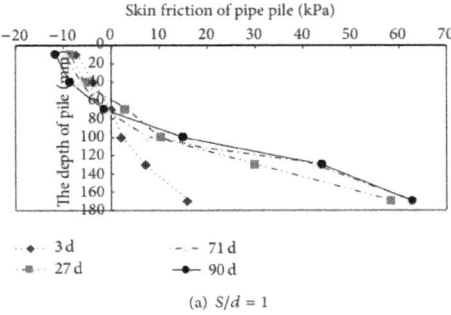

(a) $S/d = 1$

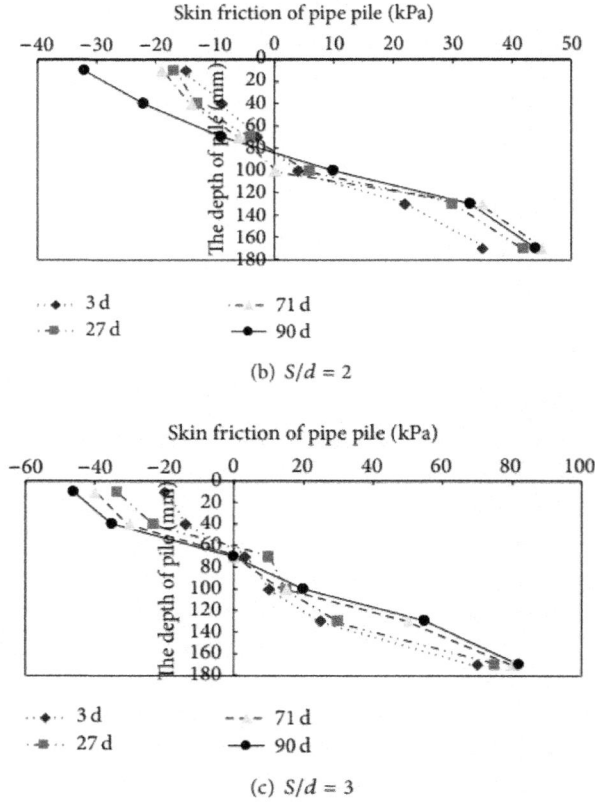

(b) $S/d = 2$

(c) $S/d = 3$

Figure 11: Skin friction distributions along the pile depth during water log infiltration.

CONCLUSIONS

The fiber Bragg grating sensor offers distinct advantages, such as its small size and limited influence on structure or over the other instruments in the skin friction monitoring of pipe piles. The model test is successfully performed using an FBG sensor that exhibits high sensitivity and resolution, even when placed in centrifugal field. Jacked pipe piles in centrifugal field can be accomplished by use of the actuator of the Chang'an University geocentrifuge. The strain variation during the pressing-in of pile was monitored by the FBG sensors, which show that the pipe pile exhibits a negative correlation between the lateral friction resistance and the ratio of S/d. The smaller the ratio, the greater the friction resistance, as the soil between the piles is more intensively compressed. Dense arrangement of piles can improve the intensity of compression, greatly reducing the collapsibility of soil.

ACKNOWLEDGMENT

This work was supported by the National Natural Science Foundation of China (Grants nos. 51008032 and 51378004).

REFERENCES

1. F. Shi, "Experimental research on load transfer mechanism of pretensioned high strength spun concrete piles," Chinese Journal of Geotechnical Engineering, vol. 26, no. 1, pp. 95–99, 2004.

2. W. Leng, L. Wei, and Z. Hua, "Experimental study on subgrade reaction under raft foundation restressed with unbonded tendon," Chinese Journal of Geotechnical Engineering, vol. 22, no. 4, pp. 456–460, 2000.

3. Y. Xiaokui, "Research on the testing of pile based on distributed optical fiber monitoring sensing technique," Electric Power Survey and Design, no. 6, pp. 12–16, 2006.

4. H.-F. Xing, H.-W. Zhao, G.-B. Ye, and C. Xu, "Analysis of engineering characteristics of PHC pipe piles,"Chinese Journal of Geotechnical Engineering, vol. 31, no. 1, pp. 36–39, 2009.

5. J. Xu, Study on the pile resistance and the optical fiber method of the jacked pile stress [Ph.D. thesis], Qingdao Technological University, Qingdao, China, 2011.

6. G. Kister, D. Winter, Y. M. Gebremichael et al., "Methodology and integrity monitoring of foundation concrete piles using Bragg grating optical fibre sensors," Engineering Structures, vol. 29, no. 9, pp. 2048–2055, 2007.

7. R. L. Idriss, M. B. Kodindouma, A. D. Kersey, and M. A. Davis, "Multiplexed Bragg grating optical fiber sensors for damage evaluation in highway bridges," Smart Materials and Structures, vol. 7, no. 2, pp. 209–216, 1998.

8. N. Metje, D. N. Chapman, C. D. F. Rogers, P. Henderson, and M. Beth, "An optical fiber sensor system for remote displacement monitoring of structures—prototype tests in the laboratory," Structural Health Monitoring, vol. 7, no. 1, pp. 51–63, 2008.

9. W. S. Zhu, Q. B. Zhang, H. H. Zhu et al., "Large-scale geomechanical model testing of an underground cavern group in a true three-dimensional (3-D) stress state," Canadian Geotechnical Journal, vol. 47, no. 9, pp. 935–946, 2010.

10. B.-J. Wang, K. Li, B. Shi, and G.-Q. Wei, "Test on application of distributed fiber optic sensing technique into soil slope monitoring," Landslides, vol.

6, no. 1, pp. 61–68, 2009.

11. X. L. Weng and W. Wang, "Influence of differential settlement on pavement structure of widened roads based on large-scale model test," Journal of Rock Mechanics and Geotechnical Engineering, vol. 3, no. 1, pp. 90–96, 2011.

12. Z. Zhou, W. Liu, Y. Huang et al., "Optical fiber Bragg grating sensor assembly for 3D strain monitoring and its case study in highway pavement," Mechanical Systems and Signal Processing, vol. 28, pp. 36–49, 2012.

13. K. O. Hill, B. Malo, F. Bilodeau, D. C. Johnson, and J. Albert, "Bragg gratings fabricated in monomode photosensitive optical fiber by UV exposure through a phase mask," Applied Physics Letters, vol. 62, no. 10, pp. 1035–1037, 1993.

14. W. W. Morey, G. Meltz, and W. H. G. Glenn, "Fiber optic bragg grating sensors," in SPIE Fiber Optic and Laser Sensors VII, vol. 1169 of Proceedings of SPIE, pp. 98–107, Boston, Mass, USA, 1989.

15. A. D. Kersey, M. A. Davis, H. J. Patrick et al., "Fiber grating sensors," Journal of Lightwave Technology, vol. 15, no. 8, pp. 1442–1463, 1997.

16. G. Esposito and H. G. B. Allersma, "Centrifuge simulation of in-situ contamination removal," inProceedings of the International Symposium on Physical Modelling and Testing in Environmental Geotechnics, pp. 141–148, La Baule-Escoublac, France, May 2000.

17. A. Y. Pasha, L. Hu, J. N. Meegoda, E. Aflaki, and J. Du, "Centrifuge modeling of in situ surfactant enhanced flushing of diesel contaminated soil," Geotechnical Testing Journal, vol. 34, no. 6, pp. 209–216, 2011.

18. G.-Q. Kong, Q. Yang, P.-Y. Zheng, and M.-T. Luan, "Model tests on negative skin friction for inclined pile considering time effect," Chinese Journal of Geotechnical Engineering, vol. 31, no. 4, pp. 617–621, 2009.

19. X. F. Huang, Z. H. Chen, S. Ha, S. G. Xue, and S. X. Sun, "Research on bearing behaviors and negative friction force for filling piles in the site of collapsible loess with big thickness," Chinese Journal of Geotechnical Engineering, vol. 29, no. 3, pp. 338–346, 2007.

Chapter 12

SETTLEMENT ANALYSIS OF SATURATED TAILINGS DAM TREATED BY CFG PILE COMPOSITE FOUNDATION

Jinxing Lai[1,2], Houquan Liu[2], Junling Qiu[2], and Jianxun Chen[1]

[1]Shaanxi Provincial Major Laboratory for Highway Bridge & Tunnel, Chang'an University, Xi'an 710064, China

[2]School of Highway, Chang'an University, Xi'an 710064, China

ABSTRACT

Cement fly ash gravel (CFG) pile composite foundation is an effective and economic foundation treatment approach, which is significant to building foundation, subgrade construction, and so forth. The present paper aims at investigating the settlement behaviors of saturated tailings dam soft ground under CFG pile composite foundation treatment, in which FEM and laboratory model test were utilized. The proposed findings demonstrate that CFG pile treatment is effective in reinforcing saturated tailings dam and loading has little influence on settlement of soil between piles. The variation of soil between piles settlement in FEM has a good agreement with the laboratory model test. Additionally, the cushion deformation modulus has a small effect on the composite foundation settlement, although the cushion thickness will generate certain influence on the settlement distribution of the composite foundation.

INTRODUCTION

With the rapid development of highway construction in China, soft ground treatment, as a limiting factor for construction schedule, costs, and engineering quality, has received more and more attentions. CFG pile composite foundation was extensively used in ground treatment due to its large extent in bearing capacity improvement, being applicable to various ground cases, rapid construction, low engineering costs, and so forth [1–5]. CFG pile is high bond strength pile formed by cement, fly ash, gravel, aggregate chips, sand, and a moderate amount of water, together with soil between piles and cushion to form composite foundation. In the 1960s, engineers applied gravel pile to consolidate natural soft ground, and thereafter it was regarded as composite foundation. Foundation Institute of China Academy of Building Research developed CFG pile composite foundation in the early 1990s [6], which greatly facilitated the development of composite foundation theory and design method. During the highway construction, CFG pile composite foundation was commonly used under flexible foundation of embankment load, in which CFG pile composite foundation was confronted with larger settlement, lack of developing of bearing capacity of pile, and unstable foundation [7–12].

To date, an increasing number of analytical, experimental, and numerical approaches have been developed to investigate the behaviors of CFG pile composite foundation under flexible foundation. In terms of theoretical analysis, for example, He et al. [13] explored the function mechanism of cushion in the CFG pile composite foundation and proposed an analytic formula of optimum pile spacing and actual replacement rate. Ji et al. [14] conducted research on load transfer mechanism of the CFG pile composite foundation on soft ground of expressways. Many researchers have carried out experimental studies to investigate the performance of CFG composite foundation. For instance, Huang [15] implemented study on the related function of CFG composite foundation under thick cushion via static load test. Zhang et al. [16] performed detailed analysis on pile-soil stress ratio with different pile spacing under embankment load through field test. Zhang et al. [17] discussed the impact of pile spacing on subgrade stability under embankment load by centrifugal model test. Numerical investigation has also been applied to study the CFG composite foundation. For example, Zheng et al. [18] used ANSYS software to investigate the behaviors of composite CFG-lime pile foundations under various load distributions. Wang et al. [19] studied the determination approach of bearing capacity of CFG pile composite foundation under the railway flexible foundation based on FEM. Zhan and Jiang [20] reported the liquefaction resistance of the CFG pile composite foundation on high-speed rail subgrade via FEM. However, studies on using CFG pile composite foundation to reinforce special soft ground of saturated tailings dam are deficient. Tailing dam is a dangerous source of artificial debris flow with high potential energy [21, 22]. Serious accidents of dam breaking should be avoided. Thus, it is of great necessity to explore the performance characteristics of saturated tailing dam used as high-level highway subgrade. Hence, the present paper employed FEM and compared the results of laboratory model test to implement detailed analysis on the settlement characteristics of the CFG pile composite foundation in saturated tailing dam under embankment load, which can provide experience to the design and construction of the CFG pile composite foundation in soft ground.

ENGINEERING OVERVIEW

Wangzhuangbao-Fanshi expressway is located in Shanxi, China; the route has nine sections drilling through or passing near saturated tailing dam area. Because the bearing capacity could not meet design requirements in the nine sections, three treatments (cushion, cement soil plus cushion, and CFG pile composite foundation) were employed in different sections, and the CFG pile composite foundation was applied at the sites between K55 + 650 and K55 + 770 sections of Yuehong Magnetic Plant saturated tailings dam (the bottom of tailing dam is "V" shaped gully, and the bottom layer is loess layer), which showed the best reinforcement effect among the three treatments. The expressway passed through the middle of tailings reservoir in the form of filling subgrade, in which CFG pile composite foundation was used to reinforce saturated tailings sand foundation (Figure 1). The length of CFG pile is 15 m and the pile diameter is 0.5 m with the pile spacing of 1.75 m. 3 m gravel cushion was filled at the bottom of the subgrade after drainage.

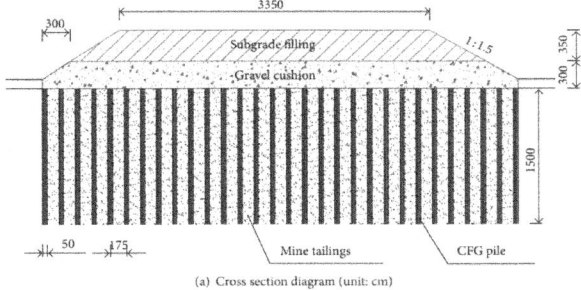

(a) Cross section diagram (unit: cm)

(b) *In situ* diagram

Figure 1: Cross section and in situ diagram of CFG pile composite foundation.

FINITE ELEMENT ANALYSIS OF CFG PILE COMPOSITE FOUNDATION

Finite Element Model and Selected Parameters

The dimensions of the FE model are of $14\,m \times 10\,m \times 31.5\,m$. Displacements are restricted at the model boundaries in the normal direction to their respective planes. From the bottom to the top of the FE model, the soil profiles consist of loess layer, saturated tailings sand layer, gravel cushion, and subgrade filling layer, respectively, in which the depths of saturated tailings sand and loess layers are 25 m in total, gravel cushion is 3 m, and subgrade filling is 3.5 m. The interface of all soil layers was simplified to be plane, and soil was assumed to be solid element which obeys elastic perfectly plastic Mohr-Coulomb yield criterion [23]. Besides, the length of CFG pile is 15 m, pile diameter is 0.5 m, and pile spacing is 1.75 m. Beam element was employed for simulation. CFG pile was simplified as linear elasticity model, and 35 CFG piles in total were simulated. Contact element was constructed between pile and soil. Uniform load was added to composite foundation, and three loadings were added during the entire simulation process. The loads used in the simulation were 50 kN/m^2, 100 kN/m^2, and 150 kN/m^2, respectively, based on the in situ loading conditions. The overall model of composite foundation is shown in Figure 2, and geotechnical properties of CFG pile and soil layers are summarized in Table 1. Soil layers parameters were acquired from the field test [24], and the CFG piles parameters were typical values [15]. Modulus of deformation was used for soil, and modulus of elasticity was employed for the CFG piles.

Table 1: Physical parameters of soils and CFG piles [24]

Number	Material types	Modulus/MPa	Poisson's ratio	Soil unit weight/(kN/m³)	Cohesion/kPa	Angle of internal friction/(°)
1	CFG pile	1600	0.25	21.5	900	35
2	Saturated tailings sand	2	0.32	18.7	7	40
3	Loess	40	0.25	20	35	45
4	Gravel cushion	140	0.16	20	0	36
5	Filling layer	100	0.2	19	15	20

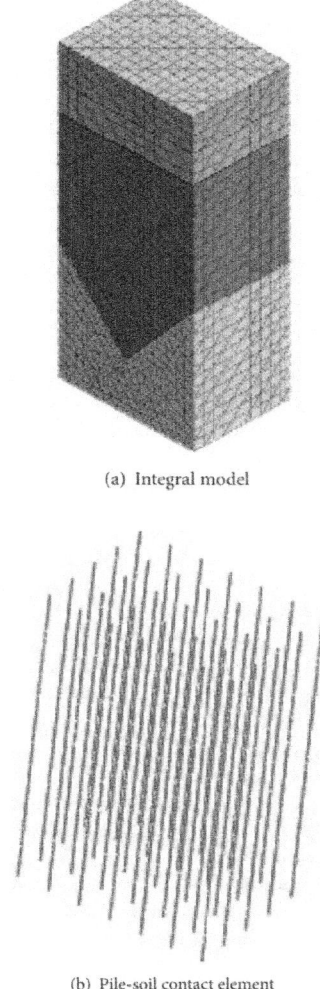

(a) Integral model

(b) Pile-soil contact element

Figure 2: FE model of composite foundation and pile-soil contact element.

FEM Analytical Results

(1) Settlement Variation with Depth. Figure 3 shows the FEM results of the whole settlement distribution of CFG pile composite foundation with depth. Figure 4 demonstrates the variation of soil layers settlement with depth. To investigate the settlement distribution of different soil layers, the central pile position of CFG pile composite foundation was selected for analysis. From the surface to the bottom of the model, totally 64 points (0.5 m interval) were chosen

for settlement analysis. Figure 4 clearly shows that similar settlement trend is found among different loadings, settlement is consistent and stable in each soil layer, and obvious variation commonly occurs to the interface of different layers. Overall, the settlement of composite foundation almost decreases with the increase of depth. However, at a depth of 3.5 m, the interface of the filling layer and gravel cushion, settlement exhibits a significant increase, and gravel cushion has the maximum settlement among the soil layers which reaches 22.4 cm when the load is 150 kPa. Hence, it seems extremely important to improve cushion material performance in the reinforcement of soft soil foundation. In the area of soil between piles (6.5–21 m depth), the settlement basically presents very small changes, which demonstrates that the saturated tailing dam has good settlement performance after reinforcement by CFG pile composite foundation. The settlement is extraordinarily small when the depth reaches 25 m and then it slowly decreases to zero with further increase of depth, which is due to the stable loess layer and reasonable pile length of 15 m in the soil layers.

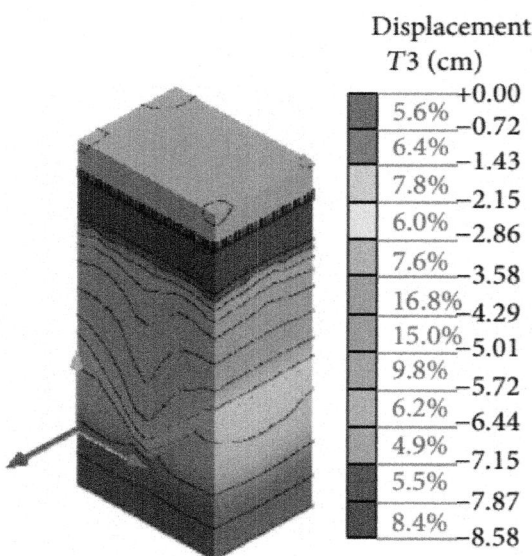

Displacement
T3 (cm)

	+0.00
5.6%	−0.72
6.4%	−1.43
7.8%	−2.15
6.0%	−2.86
7.6%	−3.58
16.8%	−4.29
15.0%	−5.01
9.8%	−5.72
6.2%	−6.44
4.9%	−7.15
5.5%	−7.87
8.4%	−8.58

(a) Case of subgrade filling construction procedure

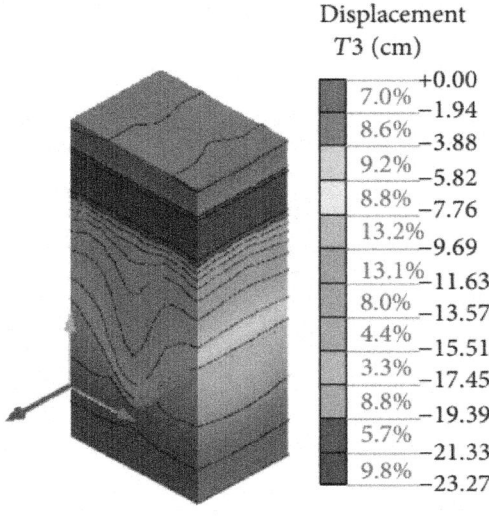

(b) Case of loading of 150 kPa

Figure 3: Distribution of soil settlement in different soil layers.

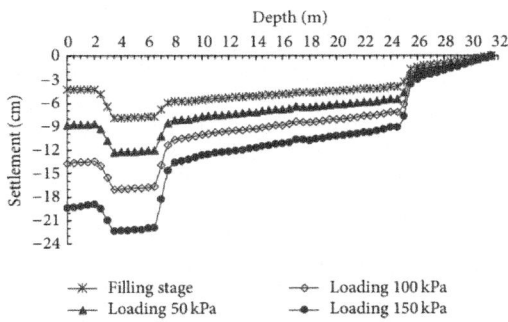

Figure 4: Variation of soils settlement with depths.

(2) Settlement Variation with Horizontal Distance. The present analysis aims at exploring horizontal variation of different soil layers settlement. Positions were selected by 0.5 m interval horizontally. For each analysis position, from the surface to the bottom of the model, settlement of the center of filling layer, center of cushion, surface of soil between piles, center of soil between piles, bottom of soil between piles, and center of loess layer were selected for analysis. The results indicate that settlement is horizontally stable, as depicted in Figures 5 and 6. Settlement in the filling layer and cushion are much stable

over horizontal distance. Settlement fluctuation is detected in the top surface of soil between piles, with amplitude of about 3.5 cm, which points out that the pile top has penetrated into cushion and resulted in differential settlement on surface of the pile top. Zhao et al. [25] conducted the study on the settlement computed by considering the effect of CFG pile top penetrating into cushion and proposed the computed method of CFG pile composite foundation settlement with the integration of composite pile-soil-cushion. Additionally, obvious "V" shaped central (central pile) settlement shift is found in the center and bottom of the soil between piles, and radius of settlement shift is about 3 m, which is consistent with the FEM CFG composite foundation settlement reported in [26, 27]. In summary, the settlement is consistent and stable relatively at subgrade transverse, which reveals that there is a favorable effect of CFG composite foundation treatment.

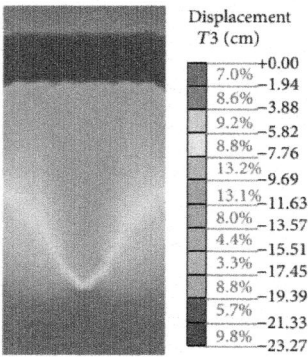

Figure 5: Variation of soil settlement in horizontal direction.

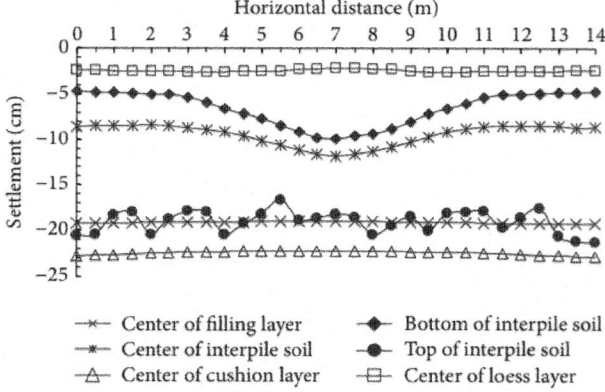

Figure 6: Variation of soil settlement with subgrade transverse distance.

(3) Settlement Variation with Load. This section aims to investigate the variation of settlement with load. Points in the central pile position (3 m interval) were selected to study settlement under load of filling construction and three different loadings. The variation of settlement for different soil layers with the load is illustrated in Figure 7. A greater increasing range of the settlement in the depth range of 0–6 m is observed, in which the increasing range at 0 m is the greatest, from 4.2 cm to 19.3 cm. This declares that it has a significant effect on the subgrade surface after loading. In the saturated tailing sand area with CFG pile treatment, the settlement increases with the increase of load with a small increasing range which is stable around 2.1 cm.

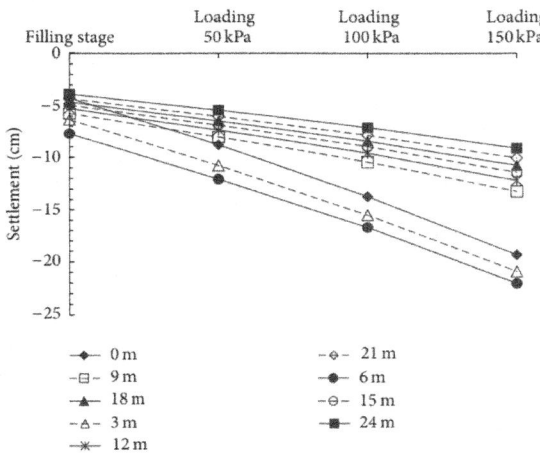

Figure 7: Variation of soils settlement under different loads.

(4) Settlement Variation with Time. The settlement in the first 15 days was analyzed by using the settlement-consolidation model. Loading was added every five days, and a total of three loadings were simulated. Settlement value was recorded each day. Settlement of the soil between piles at the central pile position (the maximum settlement) was selected for analysis. Settlement which occurred during construction phase was not taken into consideration, and only settlement happening in loading process was studied. The variation of settlement with time in composite foundation is presented in Figure 8 (where "1-2" means the settlement which was recorded at the second day after the first loading), which indicates that the settlement rate of various soil layers decreases with time. The settlement of various soil layers gradually becomes stable. In addition, after the third loading, the settlement rate of soil between piles below the depth of 9 m from 1.1 cm in the first loading reduces to 0.2 cm, and that of filling layer and gravel cushion reduces from 2 cm to 0.4 cm. The

computed settlement of soil between piles is observed to diminish at a rapid rate with time. It indicates that the saturated tailings dam presents better settlement stability after CFG pile treatment and further proves that CFG pile composite foundation obtains a favorable effect.

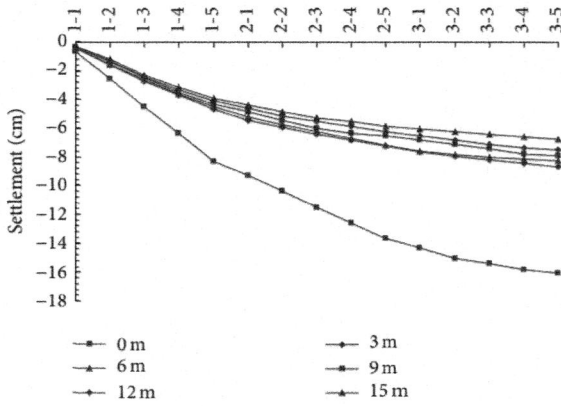

Figure 8: Variation of soil settlement with time.

LABORATORY MODEL TEST

Test Scheme

Xing [24] had conducted laboratory model test on the CFG pile composite foundation, which was based on the saturated tailings dam of Yuehong Magnetic Plant at K55 + 650–K55 + 770 sections of Wangzhuangbao-Fanshi expressway in Shanxi, China. The geometric similarity ratio a_1 of the model test is 10, and the dimension for model groove is 2 m × 2 m × 2 m. The model trough is shown in Figure 9.

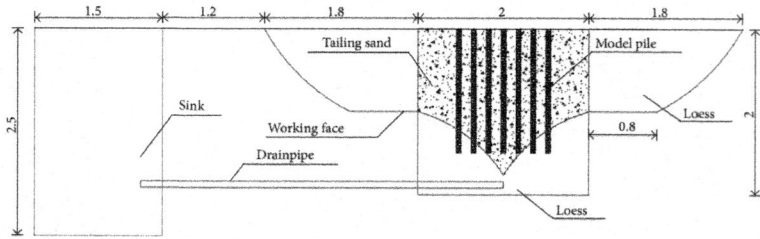

Figure 9: Laboratory model test of CFG pile (unit: m).

The CFG model pile was made of stone chip, sand, gravel, fly ash, cement, and water. The length of the CFG pile is 1.5 m, diameter is 0.05 m, and 35 CFG piles in total were employed in the model test. Layout plan of CFG pile in the model test is shown in Figure 10. The in situ picture of the model test is shown in Figure 11.

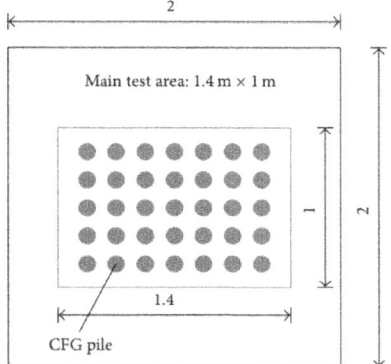

Figure 10: Layout diagram of CFG piles (unit: m).

(a) Model trough

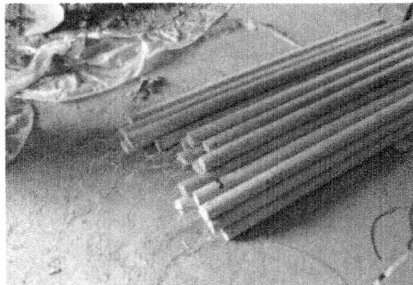

(b) CFG model pile

Figure 11: Laboratory model test model [24].

Comparison of Laboratory Model Test with FEM

The results of the FEM and laboratory model test on the CFG pile composite foundation settlement demonstrate that the settlement of various soil layers is consistent and stable at different pile positions. Therefore, the soil between piles at central pile (with maximum settlement) was selected to carry out the comparative analysis on the results between the FEM and laboratory model test which investigated the variation of composite foundation settlement with load, depth, and time. And settlement value of the laboratory model test results was multiplied by settlement similarity ratio $a_1 = 10$ according to the similarity theory.

Settlement variation of soil between piles with load is illustrated in Figure 12. The results indicate that settlement of CFG pile composite foundation under load changes evenly and finally becomes stable. The settlement increasing range in the FEM is larger than that from the laboratory model test. Obvious variation between the laboratory test and FEM modeling happens to surface of the soil between piles, which is small below the surface area. This is because FEM takes cushion and subgrade filing into account; settlement increasing range of soil between piles on the pile top surface is larger than other depths obviously.

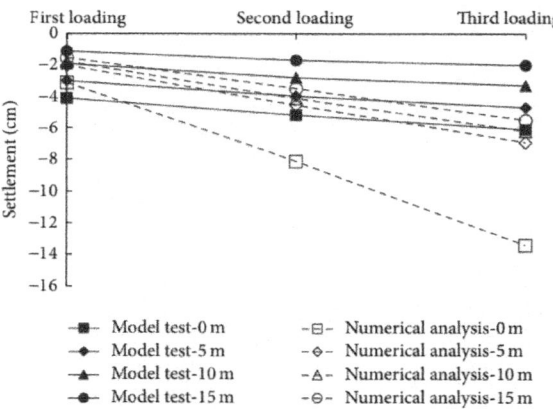

Figure 12: Comparison of interpile soil settlements under different loads between FEM and model test.

Figure 13 shows the variation of settlement of soil between piles with depth. The laboratory model test results indicate that the settlement decreasing range is relatively stable with the depth. Based on the FEM results, the settlement decreasing range is relatively large at the intermediate depths of 0–5 m, while it becomes small and tends to be stable below the depth of 5 m. Both the experimental results and FEM results show that the settlement of saturated tailing dam after CFG pile treatment exhibits even and stable changes with depth.

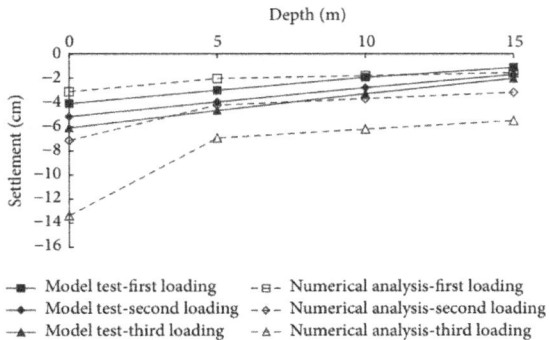

Figure 13: Comparison of interpile soil settlements at different depths between FEM and model test.

Presented in Figure 14 is the variation of settlement of soil between piles with time. The results of FEM show a good agreement with the laboratory model test. The settlement rate at different depths becomes stable in the later period. This indicates that the settlement of CFG pile composite foundation becomes stable gradually and CFG pile composite foundation has distinctive settlement characteristics and reinforcement effects.

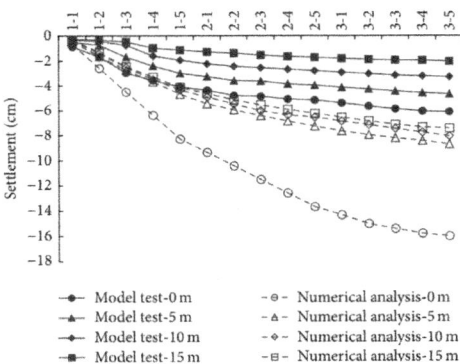

Figure 14: Comparison of interpile soil settlements with time between FEM and model test.

The variation of settlement of soil between piles in FEM exerts a good agreement with the laboratory model test; thus, to a certain extent, the FEM results can better reflect the settlement changes in practical engineering. The results from FEM analysis show that the cushion has the maximum settlement, and the average value of settlement difference between cushion and soil between piles reaches about 7 cm. Hence, it is quite important to improve cushion performance to reduce the settlement of composite foundation. And there have been some studies which discussed and introduced the effects of the cushion on the properties of composite foundation [28–30]. Both FEM and laboratory model test signify that the settlement of soil between piles tends to be consistent and stable; thus, CFG pile composite foundation has obvious treatment effect.

INFLUENCE OF CUSHION PARAMETERS ON THE SETTLEMENT OF CFG PILE COMPOSITE FOUNDATION

Pile-cushion system is important for the success of CFG pile composite foundation [15]. According to the FEM results, cushion possesses the largest settlement compared to other soil layers. Hence, it is essential to understand the influence of the cushion thickness and modulus on the CFG pile composite foundation.

The Influence of the Cushion Modulus on Settlement of the CFG Pile Composite Foundation

The deformation modulus of cushion in the aforementioned model is $E_0 = 140$ MPa. The present section aims at investigating the influence of the cushion deformation modulus on the composite foundation settlement. Settlement of the central pile position was analyzed using different values of deformation modulus of cushion (50, 100, 150, 200, 250, and 300 MPa). The settlement results from FEM analysis with $E_0 = 50$ and 300 MPa are shown in Figure 15. Composite foundation settlement at different depth under the deformation modulus of 50, 150, and 250 MPa is presented in Figure 16. Figure 17 shows the variation of the settlement with the modulus at the depths of 0, 3, and 6 m. With the increase of cushion deformation modulus, the settlement decreases at 0–5 m, but settlement increases slightly below depth of 5 m. When the deformation modulus is relatively small, increasing the deformation modulus will create a certain effect on the subgrade filling and cushion; specifically, the settlement of the two layers can be reduced by 0.8 cm when the deformation modulus increases from 50 MPa to 100 MPa. The effects on settlement become very small when the modulus is greater than 100 MPa. Additionally, improving modulus has little influence on the settlement of the saturated tailings sand. Overall, the deformation modulus of cushion around 100 MPa is reasonable.

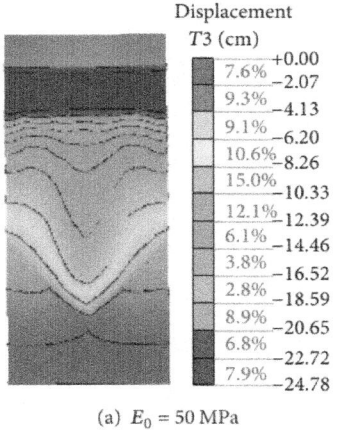

(a) $E_0 = 50$ MPa

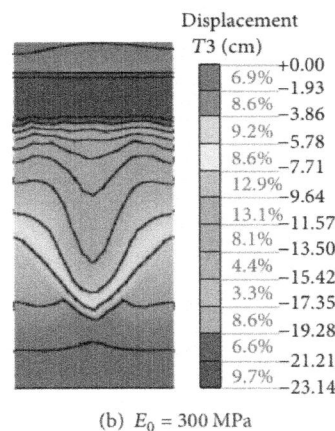

(b) $E_0 = 300$ MPa

Figure 15: Settlement distribution of composite foundation under different deformation modulus.

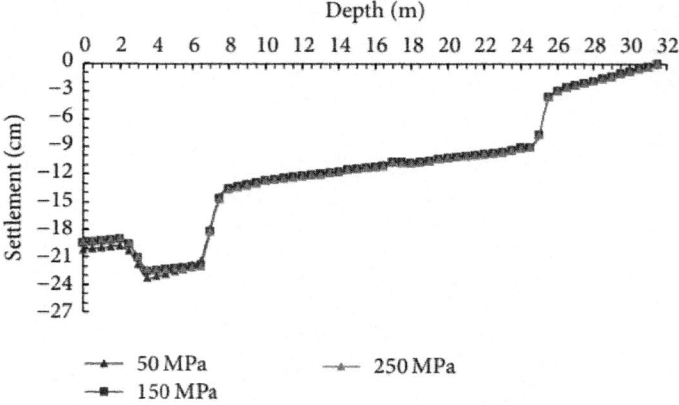

Figure 16: Distribution of composite foundation settlement under different deformation modulus with depth.

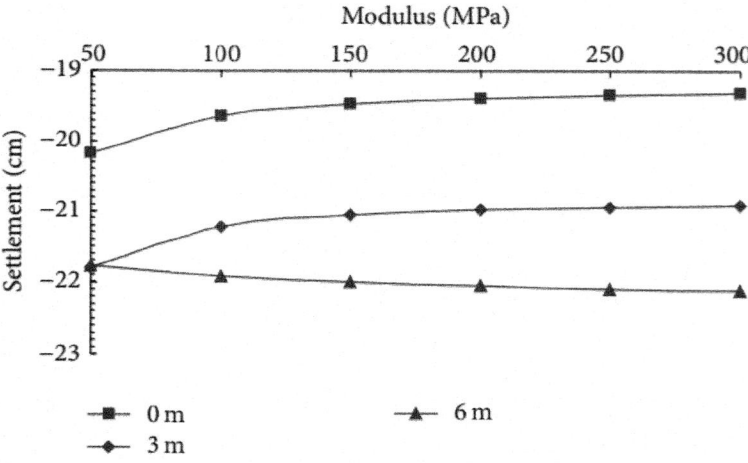

Figure 17: Variation of the settlement at central pile position under different modulus.

The Influence of the Cushion Thickness on Settlement of the CFG Pile Composite Foundation

This section aims to explore the influence of the cushion thickness on composite foundation settlement. Deformation modulus of cushion $E_0 = 140\,MPa$ was used in the elastic perfectly plastic model for the analysis while varying the thickness of the cushion (0, 0.5, 1, 1.5, and 2 m). In the FE model, the total thickness of the cushion and subgrade filling remained

6.5 m. Figure 18 shows the FEM results when the thickness of the cushion is 0.5 and 1.5 m, respectively. Distribution of composite foundation settlement under different thickness of cushion with depth is presented in Figure 19. The results (cf. Figure 19) indicate that settlement of the subgrade filling gradually decreases with the increase of the thickness when the thickness of cushion is 0–2 m. When the thickness of cushion is 0–0.5 m, subgrade filling settlement is stable, and the settlement difference is less than 2 mm. While the thickness of cushion is 0.5–1 m, settlement difference reaches 2.1 cm. Keeping increases in the thickness of the cushion, settlement appears to have a relatively consistent increase, which indicates that the thickness of cushion has great influence on subgrade filling settlement in the thickness range of 0.5–1 m. In addition, the cushion settlement increases with the increase of the thickness, and increasing range is less than 0.8 cm. In saturated tailings sand, settlement increases with the increase of the thickness of cushion. Settlement has great changes and the settlement difference reaches about 1.2 cm when thickness of the cushion is 0–0.5 m. Meanwhile, when the thickness of cushion is 1.5–2 m, settlement becomes stable. This means that thin cushion will generate great influence on the settlement of saturated tailing sand. Furthermore, when the cushion thickness is less than 1 m, settlement in the various soil layers decreases with depth gradually, and the settlement difference among the soil layers is reduced gradually; the largest settlement occurs in the surface of subgrade filling. When the thickness reaches 1 m, the reduction of settlement in different soil layers tends to be a linear change, which indicates that the settlement is very consistent. When the thickness is more than 1 m, settlement in the soil increases firstly and then decreases with depth; the largest settlement happens to the bottom of the cushion.

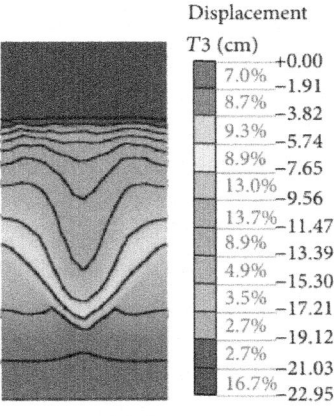

Displacement
T3 (cm)

	+0.00
7.0%	−1.91
8.7%	−3.82
9.3%	−5.74
8.9%	−7.65
13.0%	−9.56
13.7%	−11.47
8.9%	−13.39
4.9%	−15.30
3.5%	−17.21
2.7%	−19.12
2.7%	−21.03
16.7%	−22.95

(a) $d = 0.5$ m

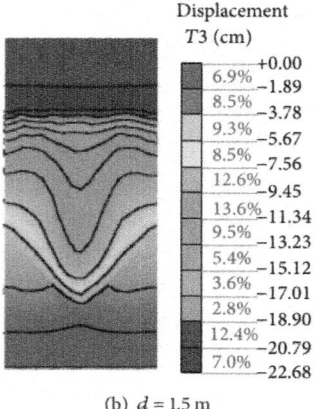

(b) $d = 1.5\,\mathrm{m}$

Figure 18: Settlement distribution of composite foundation with different thickness of cushion.

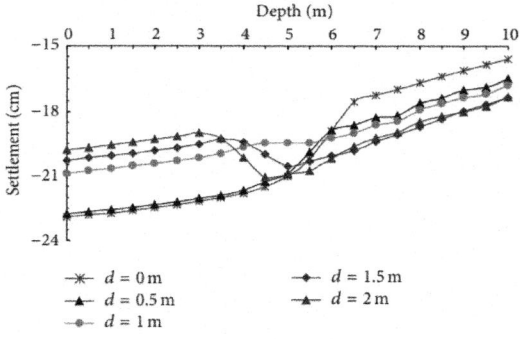

Figure 19: Distribution of composite foundation settlement under different thickness of cushion with depth.

In summary, the settlement distribution of composite foundation in different soil layers is very homogeneous when the thickness of cushion is around 1 m, while thick cushion will limit the development of the CFG pile bearing capacity. Therefore, cushion thickness of about 1 m is reasonable when CFG pile composite foundation is used in the treatment of saturated tailings dam.

CONCLUSIONS

The current paper presented FE model to study settlement behaviors of CFG pile composite foundation. In this model, the variation of the various soil layers settlement was investigated and the results were validated through comparison with the laboratory model test results. In addition, the effects of cushion on the CFG composite foundation settlement were studied. According to the results of settlement studies conducted in this research, some of the main findings are summarized as follows:

1.   According to the FEM results, it is confirmed that the distribution of various soil layers settlement is homogeneous and stable except for the interface of soil layers. After loading, settlement rate of soil between piles (saturated tailings sand) is slower than other soil layers and tends to be stable more rapidly. The increasing range of the settlement in saturated tailings sand is relatively smaller than other soil layers, which is around 2.1 cm.

2.   The variation of settlement of soil between piles in FEM is consistent with laboratory model test with depth, load, and time. Furthermore, in both cases, the results demonstrate that the various soil layers settlement of CFG pile composite foundation behavior is consistent and stable, which signifies that saturated tailing dam has obtained a good reinforcement effect after treatment of CFG pile composite foundation.

3.   When the cushion deformation modulus is around 0–100 MPa, increasing the deformation modulus will exert a certain influence on the subgrade filling and cushion. However, there is small effect on the composite foundation settlement when it is more than 100 MPa. The thickness of cushion has a relatively large influence on the settlement distribution of the composite foundation. When the thickness of cushion is around 1 m, the settlement has approximately linear decrease with depth. At the same time, the settlement distribution of composite foundation in various layers is more homogeneous than the case of the other cushion thickness. Therefore, it is suggested that modulus of deformation of about 100 MPa and cushion thickness of about 1 m are reasonable, respectively.

4.   Based on the results of FEM, cushion has the maximum settlement at different loading cases which is 7 cm larger than that of soil between piles on average. In addition, the settlement control effects are not quite obvious by means of changing the modulus and thickness of the cushion. Therefore, in order to further optimize the design scheme for CFG pile composite foundation and improve the performance characteristics of composite foundation, further experimental studies should be conducted on the new cushion material and the novel connecting

type among cushion, pile, and soil of CFG pile composite foundation under flexible foundation.

ACKNOWLEDGMENTS

This work is financially supported by the Special Fund for Basic Scientific Research of Central Colleges of Chang'an University (Grant no. 31082116011), the Key Industrial Research Project of Shaanxi Provincial Science and Technology Department (Grant no. 2015GY185), and the Integrated Innovation Project of Shaanxi Provincial Science and Technology Department (Grant no. 2015KTZDGY01-05-02).

REFERENCES

1.   A.-J. Zhou and B. Li, "Experimental study and finite element analysis of cushion in CFG pile composite foundation," Rock and Soil Mechanics, vol. 31, no. 6, pp. 1803–1808, 2010.

2.   X. Pan, "Study of settlement analysis of composite foundation of CFG piles," Rock and Soil Mechanics, vol. 26, pp. 248–251, 2005.

3.   M. L. Yan, C. L. Wu, and J. Yang, "Study on the composite foundation with cement-flyash-gravel pile,"Chinese Journal of Geotechnical Engineering, vol. 18, no. 2, pp. 56–62, 1996.

4.   F. Song, J. M. Zhang, and G. R. Cao, "Experimental investigation of asymptotic state for anisotropic sand," Acta Geotechnica, vol. 10, no. 5, pp. 571–585, 2015.

5.   X. Weng, Y. Nie, and J. Lu, "Strain monitoring of widening cement concrete pavement subjected to differential settlement of foundation," Journal of Sensors, vol. 2015, Article ID 679549, 7 pages, 2015.

6.   L. Zhao, X. H. Wang, and H. B. Wu, "Settlement calculation of CFG pile composite foundation under flexible foundation," Subgrade Engineering, vol. 3, no. 138, pp. 147–149, 2008.

7.   Y. Zhang and Q. Huang, "Analysis on the properties of semi rigid pile composite foundation," Journal of Geotechnical Engineering, vol. 15, no. 2, pp. 86–93, 1993.

8.   J. W. Duan, X. N. Gong, and G. X. Zeng, "Load transfer behavior of cement treated soil column," Chinese Journal of Geotechnical Engineering, vol. 16, no. 4, pp. 2–7, 1994.

9.   J. Y. Chi, E. X. Song, and Z. Y. Chen, "Proportion of loads carried by the piles and the soil of rigid pile composite-foundation under varying loads," Journal of Tianjin University, vol. 36, no. 3, pp. 359–363, 2003.

10. M. Zhou, W. C. Yuan, and Y. Zhang, "Seismic material properties of reinforced concrete and steel casing composite concrete in elevated pile-group foundation," Polish Maritime Research, vol. 22, supplement 1, pp. 141–148, 2015.

11. C. G. Zhang, X. D. Chen, W. Fan, and J. Zhao, "A new unified failure criterion for unsaturated soils,"Environmental Earth Sciences, vol. 74, no. 4, pp. 3345–3356, 2015.

12. J. Lai, J. Qiu, Z. Feng, J. Chen, and H. Fan, "Prediction of soil deformation in tunnelling using artificial neural networks," Computational Intelligence and Neuroscience, vol. 2016, Article ID 6708183, 15 pages, 2016.

13. J.-B. He, B.-N. Hong, and G.-F. Qiu, "Research on cushion action mechanism of CFG pile composite foundation for expressway," Rock and Soil Mechanics, vol. 25, no. 10, pp. 1664–1666, 2004.

14. P. Ji, Y.-Q. Zou, S.-Y. Liu, X.-Y. Zhu, and D.-M. Ma, "Load transfer mechanism of cement-flyash-gravel pile in soft clay by high strain testing," Rock and Soil Mechanics, vol. 26, supplement, pp. 69–72, 2005.·

15. S. G. Huang, "Test study and finite element analysis of CFG composite foundation," Rock and Soil Mechanics, vol. 29, no. 5, pp. 1276–1279, 2008.

16. H. Zhang, Y. P. Liu, J. H. Li, and Y. Zhou, "In situ test on pile-soil stress ratio of CFG pile composite foundation under embankment load," in Proceedings of the 3rd International Conference on Transportation Engineering (ICTE ‹11), pp. 1511–1516, ASCE, Chengdu, China, July 2011.

17. J. S. Zhang, C. Guo, and S. W. Xiao, "Analysis of effect of CFG pile composite foundation pile spacing on embankment stability based on centrifugal model tests," Applied Mechanics and Materials, vol. 178, no. 181, pp. 1641–1648, 2012.

18. J.-J. Zheng, S. W. Abusharar, and X.-Z. Wang, "Three-dimensional nonlinear finite element modeling of composite foundation formed by CFG-lime piles," Computers and Geotechnics, vol. 35, no. 4, pp. 637–643, 2008.

19. L. J. Wang, G. L. Ding, S. C. Liu, X. Chen, and Y. Shen, "Study on the determination method for the bearing capacity of CFG pile composite foundation under the flexible foundation in railway," China Railway Science, vol. 29, no. 6, pp. 13–17, 2008.

20. Y. X. Zhan and G. L. Jiang, "Numerical simulation of the liquefaction resistance characteristics of CFG pile column-net composite

foundation," China Railway Science, vol. 29, no. 5, pp. 1–6, 2008.

21. R. Chen, I. Lee, and L. Y. Zhang, "Biopolymer stabilization of mine tailings for dust control," Journal of Geotechnical and Geoenvironmental Engineering, vol. 141, no. 2, Article ID 04014100, 2015.

22. R. Chen, L. Y. Zhang, and M. Budhu, "Biopolymer stabilization of mine tailings," Journal of Geotechnical and Geoenvironmental Engineering, vol. 139, no. 10, pp. 1802–1807, 2013.

23. H. Jiang, "Failure criteria for cohesive-frictional materials based on Mohr-Coulomb failure function,"International Journal for Numerical and Analytical Methods in Geomechanics, vol. 39, no. 13, pp. 1471–1482, 2015.

24. Y. R. Xing, The model test of saturated tailing ore foundation which treated by CFG piles [Ph.D. thesis], Chang›an University, Xi›an, China, 2014.

25. M.-H. Zhao, L.-P. He, and L. Zhang, "Settlement calculation of CFG pile composite foundation based on load transfer method," Rock and Soil Mechanics, vol. 31, no. 3, pp. 840–844, 2010.

26. G. L. Ding, S. C. Liu, and L. J. Wang, "Analysis of influence of construction phase on settlement and stress of CFG pile composite foundation," Highway, no. 10, pp. 13–17, 2011.

27. L. M. Xie, "Numerical analysis on the settlement of CFG composite foundation in Wuhan-Guangzhou special passenger line," Journal of Wuhan University of Technology (Transportation Science & Engineering), vol. 35, no. 6, pp. 1174–1177, 2011.

28. J. T. Shahu, M. R. Madhav, and S. Hayashi, "Analysis of soft ground-granular pile-granular mat system,"Computers and Geotechnics, vol. 27, no. 1, pp. 45–62, 2000.

29. H. F. Schweiger and G. N. Pande, "Numerical analysis of stone column supported foundations,"Computers and Geotechnics, vol. 2, no. 6, pp. 347–372, 1986.

30. F.-Y. Liang, L.-Z. Chen, and X.-G. Shi, "Numerical analysis of composite piled raft with cushion subjected to vertical load," Computers and Geotechnics, vol. 30, no. 6, pp. 443–453, 2003.

Chapter 13

VALIDATING AND IMPROVING MODELS FOR VIBRATORY INSTALLATION OF STEEL SHEET PILES WITH FIELD OBSERVATIONS

A. M. J. Mens[1, 2], M. Korff[1, 3], A. F. van Tol[1, 2]

[1]Deltares, Unit Geo-Engineering, 2600 MH Delft, The Netherlands

[2]Faculty of Civil Engineering and Earth Sciences, Delft University of Technology, Delft, The Netherlands

[3]Cambridge University, Cambridge, UK

ABSTRACT

Vibratory driving is the most common installation technique for steel sheet pile walls. In practice, the assessment of the feasibility of this installation process is mainly based on rules of thumb, on numerical and empirical models or on experts opinions. In order to improve these prediction methods and formulas, 252 observations from the Dutch engineering practice have been compared with six different types of models. This comparison has been carried out applying the receiver operating characteristic (ROC) curve technique, which is new in geotechnical engineering. This paper introduces the ROC-curve technique to estimate mainly the quality of a model and to be able to optimize parameters and variables in the model. 252 field observations were used to re-examine prediction methods for the minimum required vibration force and to prove the ROC method works. The paper shows this technique is suitable for three purposes: (1) determining the quality of a model, (2) objectively comparing several models to each other, given certain assumptions and (3) for optimizing thresholds within a model. The model with added professionals' experience proves to perform equally well as the numerical model Hypervib-I.

INTRODUCTION

Steel sheet pile walls often support deep excavations in urban areas. The most common installation technique for steel sheet pile walls is vibratory driving. Most projects are carried out in areas where the subsoil consist of several soft clay and peat layers on top of a medium or loosely compacted sand layer. Vibratory driving is attractive in such sub-soils, because of the relative straightforward technique and the high production rates. In practice, the feasibility of the installation process of these sheet piles is mainly based on rules of thumb, on empirical models or experts opinions. It is to be expected that the more experience is added to the rules and models, the more reliable prediction models become.

In 2004 Van Baars used the results of 18 observations to show the inferior quality of several models that predict the minimum required vibration force in order to determine the best vibrator for pile installations (Van Baars 2004). This current paper will show a single error is not sufficient to determine the quality of a model. The method of 'Receiver Operating Characteristic' (ROC) Curves (Metz 1978) is introduced to overcome this problem. Instead of 18 observations, 252 observations were used to re-examine the methods described by Van Baars. A positive side effect of the ROC method comprises the possibility of threshold and parameter optimization, which will be shown as well.

This paper first describes the origin of the (Dutch) field observations that were used for the comparison of the design codes (Sect. 2). Subsequently, Sect. 3 shortly describes the prediction models and the parameter choices. Section 4 introduces the essentials of the ROC-curve technique and finally the results-section will compare the models, based on the field observations, using the ROC-curve technique.

DATA–OBSERVATIONS

In order to validate the prediction models for vibratory driving, 252 field cases from the Dutch experience database 'GeoBrain' have been used (www.geobrain. nl). The GeoBrain experience database (Barends 2005; Hemmen 2005) contains case histories of foundation techniques. As most cases are unique, the experience gathered focuses on the generic techniques applied in these cases, such as piling. Since 2005, different contractors have been filling this database with their up to date experiences in the Netherlands. The total number of entries counted 2900 projects by the end of 2011. At the time of this evaluation in February 2009, 364 entries concerned the vibratory installation of steel sheet piles. An 'experience' or 'observation' is uniquely defined by the type of element (for example sheet-pile or prefabricated concrete pile), the type of equipment used and the soil conditions present. Additionally to this digitalized data, also details concerning the building pit, the crew and the surroundings have been included.

Although the database comprised 364 observations for vibratory driving at the time, only 252 of them have been used for this evaluation. An observation was discarded when

- essential data was lacking (like a Cone Penetration Test);

- a combination of installation techniques was used (both hammering and driving);

- unexpected obstacles were expected or detected in the subsoil;

- erroneous data was recorded, e.g. large differences (>1.5 m) between the entered length of the sheet pile and the difference between the head and the toe of the pile

- the head of the pile was deeper than 1.5 m below the surface.

A detailed example of one observation has been described in Mens and van Tol (2010).

In general, the projects from the GeoBrain database comprise, amongst others, the following features: the type of the vibration equipment, the type of pile that has been used, the results of one cone penetration test that reflects the mean circumstances and the number of piles that was used. "Appendix 1" shows eight boxplots with the frequency, displacement amplitude, eccentric moment, mass of the sheet pile, dynamical mass of sheet pile and vibrator, the pile length, the pile cross sectional area and the number of piles that were used for this investigation. More information about the projects and the geological area is available at geobrain.nl.

An observation is defined to be 'positive' whenever within the project 100 % of the piles reach the pre-determined depth. This 100 % avoids major subjectivity, but is of course quite conservative. A short description of each design tool follows, together with a transformation to one single criterion that makes comparison between the methods possible.

PREDICTION MODELS

Current (European) practice uses four categories of models to predict the vibratory driving equipment for successful installation of a steel sheet pile, depending on the scope and the complexity of the project. Category one comprises (numerical) computer models, such as Hypervib-I (Holeyman and Legrand 1994; Holeyman et al. 1996; Holeyman et al. 2002; Gonin et al. 2006), the Karlsruhe model (Dierssens 1994) and Vipere (Vanden Berghe 2001). Viking (2002) presents an extensive explanation and comparison of these methods.

This study regards Hypervib-I, since this computer model is in use in this region and it has been the basis for a simplified prediction equation used in the Netherlands. The second category comprises design equations and empirical rules. Some of these are in fact simplifications of more complex computer models, adapted to specific circumstances. This study discusses a rule of thumb from the CUR [the Dutch centre for research, rules and regulations in the civil engineering practice (CUR166 2005)], the EAU (the German equivalent) (EAU 1990) and two design rules based on Hypervib-I (Azzouzi 2003; Van Baars 2004). (The term Vibrive is an equivalent to Hypervib-I, Viking (2002) mistakenly added a 'd' to Vibrive, making it Vibdrive, which has been quoted as such by Van Baars (2004)). The third category comprises design charts, developed by the NVAF (the Dutch federation of foundation contractors (Van Baars 2004; CUR166 2005). The fourth and last category contains a Bayesian Belief Model, based on expert knowledge (Bles et al. 2003; Hemmen 2005) Since this is neither a numerical model, nor a simple design equation or chart, it does not belong in one of the previous categories.

Objective Criteria

Engineers need an objective criterion to compare predictions to real results for each observation. In this research the percentage of refusals, piles in a project that did not reach the predetermined depth, is taken as criterion. A project has been defined as a (part of a) construction work that uses the same equipment and the same pile-type, in an area that can be described by one 'representative' Cone Penetration Test (CPT). All prediction tools can be 're-arranged' to predict whether the combination of soil, equipment and piles is sufficient to reach the pre-determined depth. In this paper, a 'positive' prediction is equivalent to 'technically being able to reach the pre-determined depth with the chosen equipment'.

Methods

Method 1 (Hypervib-I)

The computer model Hypervib-I (Holeyman et al. 1999) regards the sheet pile as a rigid body and models the vibratory installation of the pile, making use of four forces. (1) the vertical vibration force from the vibrator on top of the pile, (2) the resisting force on the shaft from the soil friction (3) the resisting force on the tip of the pile caused by the soil in the downward motion and (4) the gravitation forces on the total mass of the system. The model includes a strength reduction by degradation or liquefaction, both at the shaft and the tip of the pile. Using Newton's second law of motion, the model provides a velocity profile, based on a cone penetration test (CPT) and the specifications of both the sheet pile and the driving equipment. The occurrence of zero velocity equals refusal and is the criterion for not reaching the pre-determined depth. The original computer program bases its prediction on the time to penetrate 1 m of soil (1/velocity). Exceeding 999 s (>16 min.) means refusal and leads to a 'negative' prediction. Therefore, this study uses a threshold value of $V_t = 1e{-3}$ m/s (6 cm/min.) for the velocity profile. The parameters in the Hypervib-I code have been determined, based on Belgium engineering practice.

Method 2 (CUR)

The 'CUR-rule' calculates the free displacement amplitude (d), that is used to determine the appropriate vibration equipment (CUR166 2005):

$$d = \frac{M_e}{m_v + m_p}$$

(1)

where d = the displacement amplitude (m); M_e = the eccentric moment (kg m); m_v = the vibrating mass of the vibrator (kg) and m_p = the mass of the sheet pile (kg).

The vibrator to be chosen must have M_e large enough to fulfill the required displacement amplitude (larger than 0.005 m). If so the sheet piles will reach the pre-determined depth. To obtain the least required cyclic force, the eccentric moment (M_e) is multiplied with $(2\varpi f)^2$, where f denotes the frequency of the equipment in Herz.

Remarkably, there is no soil involved in this equation. The idea behind this simple rule of thumb is that with an amplitude of 5 mm the sheet pile is able to degenerate the strength of the soil to relatively low values and as consequence the original strength of the soil is not involved any more.

Method 3 (Azzouzi)

Based on the Hypervib-I model (Holeyman et al. 2002), Azzouzi (2003) developed a formula that calculates the required vertical cyclic force (Fc) to be able to determine the most suitable vibrator. Azzouzi used 180 calculations with the Hypervib1 model for the development of the formula that uses the mean cone resistance over the considered sand layers, taken from a cone penetration test (CPT):

$$F_{c,\text{Azzouzi}} = \alpha_A \cdot L \cdot \chi \cdot f_A(q_c) + \beta_A \cdot A_t \cdot g_A(q_c)$$

(2)

where $F_{c,\text{Azzouzi}}$ = the required vertical cyclic force from the vibrator that should be used (N); L = pile penetration length (m); χ = the perimeter of the sheet pile (m); $f_A(q_c) = q_{c,\mathrm{mean}}$ = the mean cone resistance over the sand layers (N/m^2); $g_A(q_c) = q_{c,\mathrm{tip}}$ = the cone resistance at the tip of the pile (N/m^2); A_t = the cross-sectional area of the toe of the sheet pile in m^2; $\alpha_A = 1.92 \times 10^3 (-)$ and $\beta_A = 1.2 \times 10^{-2}(-)$.

(Formula and unities adapted to the standard SI system).

When the used force is larger than the required force, the prediction counts for 'positive'; and otherwise it is a 'negative' prediction.

Method 4 (Van Baars)

Van Baars (2004) developed a slightly different formula, based on the same 180 calculations:

$$F_{c,\mathrm{van\,Baars}} = \alpha_B \cdot L \cdot \chi \cdot f_B(q_c) + \beta_B \cdot L \cdot A_t \cdot g_B(q_c) \tag{3}$$

where $F_{c,\mathrm{van\,Baars}}$ = the required vertical cyclic force from the vibrator that should be used in N; $f_B(q_c) = g_A(q_c) = q_{c,\mathrm{tip}}$; $g_B(q_c) = \exp(\gamma \cdot q_{c,\mathrm{tip}})$; $\alpha_B = 1.2(-)$; $\beta_B = 26.4 \times 10^{-3}(-)$, and $\gamma = q_{c,\mathrm{ref}} = 8.7$ N/m^2.

Again, formula and unities adapted to the SI-system and when the used force is larger than the required force, the prediction counts for 'positive'; and otherwise it is a 'negative' prediction.

Method 5 (EAU)

The German design rule (EAU 1990) calculates the minimum required vertical force using only the length of the pile and the dynamical mass m_d

$$F_{c,\mathrm{EAU}} = \alpha_E \cdot L + \beta_E \cdot m_d \tag{4}$$

where $F_{c,\mathrm{EAU}}$ = the required vertical cyclic force from the vibrator that should be used (N); m_d = the dynamical part of the vibrator and the pile (kg); $\alpha_E = 15 \times 10^{-3}$(N/m) and $\beta_E = 3 \times 10^{-4}$(N/kg).

This rule quantifies the German criteria for their choice of vibratory equipment. For each meter of the pile one requires at least 15 kN vertical cyclic force and additionally for each 100 kg dynamical mass one needs 30 kN vertical cyclic force.

When the used force is larger than the required force, the prediction counts for 'positive'; and otherwise it is a 'negative' prediction.

Method 6 (Bayesian Belief Network)

Bles et al. (2003) developed a Bayesian Belief Network (BBN), based on professionals' experience (also abbreviated as 'experts'), to model the risks during installation of foundations. In general, BBNs use probabilistic theory for reasoning under uncertainty and risk in expert systems. Bayes' theorem is the cornerstone in this way of reasoning, because it provides a way to calculate the posterior probability. Bayes calculates the probability on some hypothesis h, given condition D: $P(h|D)$. This conditional probability of h, given D, is calculated using the prior probability $P(h)$, together with the probability on the (data-based) evidence $P(D)$ and the probability on the data, given the stated hypothesis $P(D|h)$ (Mitchell 1997):

$$P(h|D) = \frac{P(D|h)P(h)}{P(D)}$$

(5)

The method transforms joint probability functions to a set of stochastic variables, ordered in a network. The network itself consists of two parts. The qualitative part shows the relations between the variables in a graphical representation (the network). The quantitative part assigns conditional probabilities to all variables, using likelihood-tables, which describe the effect of preceding variables on the underlying ones.

The input variables include information about the subsurface (cone penetration test, presence of stiff clay or gravel, ground water level, etc.), the sheet pile (length, type, profile, mass, shape, etc.) and the method of installation (equipment, force, etc.). Experts from the Dutch Association for Contractors in Foundation Engineering (NVAF) supplied the necessary information for the likelihood tables, describing the qualitative part of the BBN. Finally, the BBN provides the user with a number between 0 and 100, describing the expected amount of risk. The lower the number, the smaller the expected problems, involving not reaching the pre-determined depth. Another study (Mens et al. 2008) suggests a threshold value of 38 %, above which to start getting worried about the risks. Above this number the prediction is considered to be 'negative' and below or equal to, it is considered to be 'positive'.

ROC-THEORY

How to measure the quality of a prediction? The diagnostic 'accuracy'—the fraction of cases for which the prediction appeared to be correct- can be very misleading. In case of binary predictions there are four possibilities: a positive prediction, that in reality fails (1) is a false positive (FP). A positive prediction that in reality is a success (2) is a true positive (TP). On the other hand, a negative prediction, carried out anyway and leading to a positive observation is false negative (FN), where as it fails as predicted it is called a true negative (TN). Now suppose only 5 % of the cases will not reach the predetermined depth and a model predicts the drivability to be always possible, its accuracy will be 95 %. But, this is based on 0 true negatives ánd 0 false negatives.

This paper introduces the 'receiver operating characteristic' (ROC)-curve technique within foundation engineering to provide a better criterion. The technique itself is not new and has been used extensively in other scientific areas, such as medical science (Metz 1978) and ecological engineering (Fawcett 2006).

Basically, this technique labels a method, based on both the sensitivity (the number of true positive predictions/the total number of positive predictions, see Eq. 6) and the specificity (the true negative predictions/the total number of negative predictions). Both of these performance indicators make up a 'sensitivity-pair', which can be plotted in a so called 'ROC-space', with the sensitivity on the vertical axis and (1 − specificity) on the horizontal axis (Figs. 1, 2, 3, 4, 5). By visualizing the sensitivity-pairs for all the mentioned design methods in one graph, an objective comparison between the tools is possible. Metz (1978) and Fawcett (2006) explain this theory in more detail. The sensitivity-pair (0,1) (Figs. 1, 2, 3, 4, 5) describes the 'perfect' model. The closer a random sensitivity pair is to this perfect point, the better the model or design rule is.

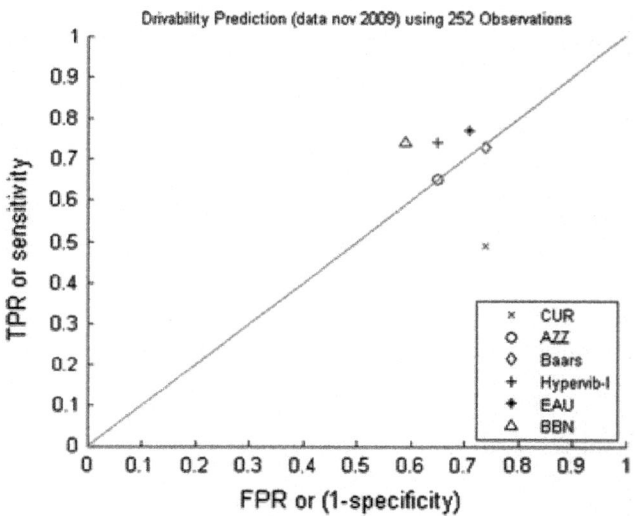

Figure 1: Drivability prediction as a sensitivity-pair for 6 design codes, using 252 field observations. *TPR* true positive ratio, *FPR* false positive ratio.

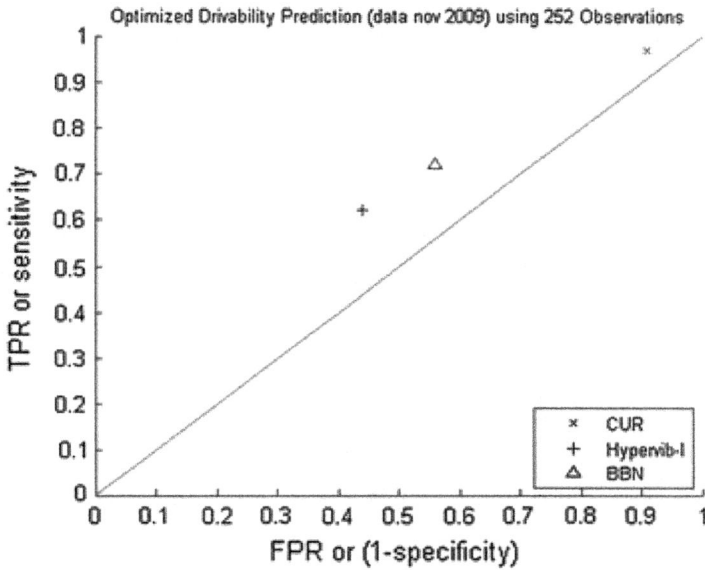

Figure 2: Optimized driveability prediction for BBN, CUR and Hypervib-I code, changing the threshold values as indicated in Sect. 5.2.

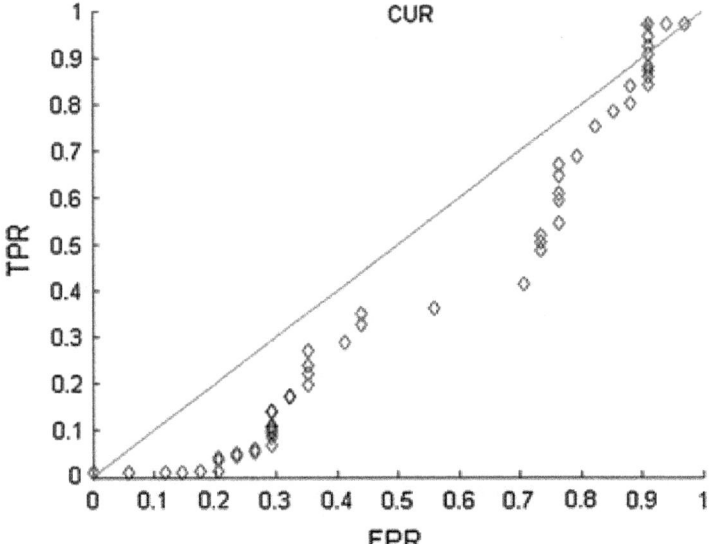

Figure 3: ROC-curve for CUR rule, where displacement amplitude *d* varies between 0 and 0.01 m. The *red star*shows the 'best' sensitivity pair, for *d* = 0.0029, assuming both the TPR and the FPR are equally important.

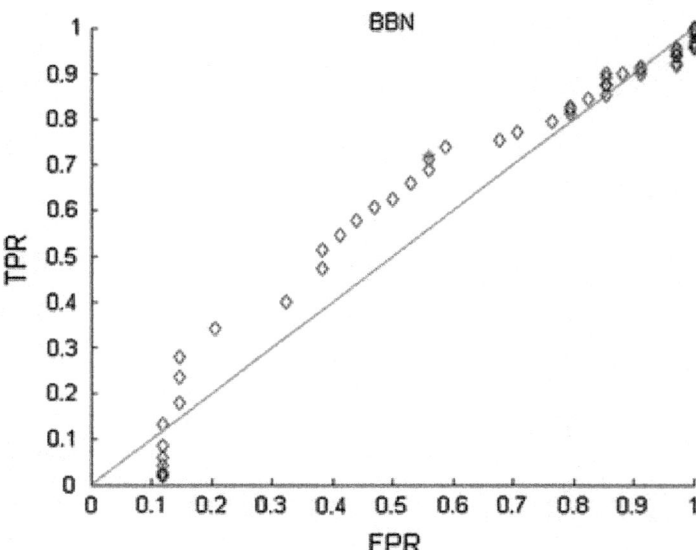

Figure 4: ROC-curve for the BBN model, where displacement amplitude *d* varies between 0 and 100 %. The *red star* shows the 'best' sensitivity pair, for threshold value = 36 %, assuming both the TPR and the FPR are equally important.

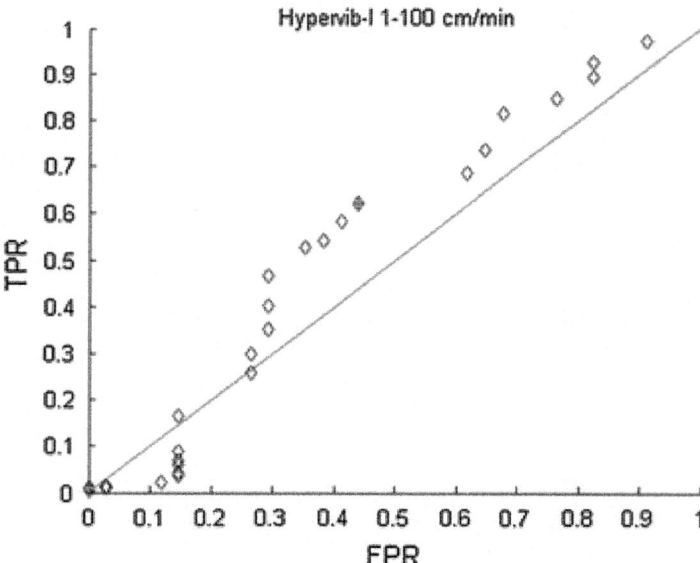

Figure 5: ROC-curve for the Hypervib-I model, where velocity threshold Vt varies between 1 and 100 cm/min (see Sect. 3.2.1). The *red star* shows the 'best' sensitivity pair, for threshold value = 8 cm/min, assuming both the TPR and the FPR are equally important.

ROC GRAPH AND CONTINGENCY TABLE

The input for the ROC graph is a given prediction model and a set with N observations (or field experiences). For these N observations, the binary result (positive or negative) is known. Using the information from these observations, it is possible to calculate the binary prediction results. These can be summarized in a two-by-two contingency table (or 'confusion'-matrix), which serves as the base for a point in the ROC-space. Table 1 provides an example of this matrix. $O(-)$ represents the total number of negative observations and $O(+)$ the positive ones. $P(-)$ and $P(+)$ represent the total number of negative and positive predictions respectively.

Table 1: Example of a contingency table, or confusion matrix

	Predictions		
	−	+	**Total**
Obs.			
−	TN	FP	$O(-)$
+	FN	TP	$O(+)$
Total	$P(-)$	$P(+)$	N

The numbers from Table 1 enable calculating the following characteristics (amongst others):

$$\text{Sensitivity} = \text{TPR} = \frac{\text{TP}}{O(+)} \tag{6}$$

$$1 - \text{Specificity} = \text{FPR} = \frac{\text{FP}}{O(-)} \tag{7}$$

These characteristics are depend on the threshold value used in the model. Take the Bayesian Belief Network (BBN) prediction model as an example. The BBN predicts a project to contain more unwanted events, if the resulting number (on a scale from 0 to 100 %) exceeds the threshold value, 38 % in this case. This 38 % determines more or less the result of the contingency table. Obviously, if we take 38 % as a threshold value, our contingency table will be different than if we take 50 % for a threshold. Table 2 shows the contingency table for the CUR-model, using the previously mentioned 252 observations from the GeoBrain experiences database.

Table 2: Example of a completed contingency table, or confusion matrix for the CUR rule, using threshold$d$ = 0.005 m

	Predictions		
	−	+	**Total**
Obs.			
−	TN = 9	FP = 25	O(−) = 34
+	FN = 111	TP = 107	O(+) = 218
Total	P(−) = 120	P(+) = 132	N = 252

Sensitivity Pair

Fortunately, a point in the ROC-space incorporates this threshold in its graph and therefore it is in fact an elaboration on the contingency table. The ROC point uses the fact that the true negative ratio (TNR(=TN/O(−))) plus the FPR equals 1, just like the TPR plus the false negative ratio (FNR(=FN/O(+))) equals 1. See also Eqs. 6 and 7. For different threshold values it is now possible to calculate the so-called sensitivity-pair (TPR,FPR). Fawcett (2006) explains this in more detail. The smaller the metric distance to the 'perfect model' (coordinates (0,1) in the ROC-space), the better the model is, assuming the TPR and the FPR are equally important. In practice, this is not correct because the costs of FP's are higher than for FN's. The effect of assigning different weights to FP's and FN's is studied in Sect. 5.2.1.

Using Table 2, the TPR equals 107/218 = 0.49 and the FPR equals 25/34 = 0.74 for the CUR rule. Therefore the sensitivity-pair (FPR, TPR) reads (0.74,0.49). Figure 1 shows this as a cross in the lower right corner. Under the assumptions that (1) the probability of a positive observation is approximately 95 % (P(O+) = 0.95) and that (2) the model predicted a positive result (P(P+) = 1), Bayes' rule (see Sect. 3.2.6) translates the true positive ratio (TPR = P(P(+)|O(+))) into a probability of failure1 P(O(−)|P(+)) = 0.53.

Extension to ROC Curves

An additional advantage of this method is the possibility of creating ROC-curves. By changing the threshold value in a design rule, the sensitivity-pair will change as well. So, for a range of threshold values, a range of sensitivity-pairs can be constructed, resulting in a ROC –curve. One point of this curve will be closest to the perfect model and this leads to the optimal threshold value. Figures 3, 4, 5 show examples for such curves, demonstrating the behavior of the CUR-rule, the BBN prediction model and the Hypervib-I model respectively.

Conservative Predictions

Whenever a sensitivity pair is located at the lower left corner of the ROC-space, one can call the corresponding model 'conservative'. An example contingency table explains why (Table 3). This table provides a FPR of 0.12 and a TPR of 0.40, creating a sensitivity pair in the lower left corner of the ROC-space. The underlying model is better than a random guess, because TPR > FPR. Furthermore, the table shows a large amount of negative predictions (161), although in reality there were 218 positive observations. The model seems to perform very well, after all, given a positive prediction 96 % of the cases provides a positive observation. You might argue that a negative prediction will lead to a new design and therefore the project will not be carried out. Reality though proves otherwise: 161 negative predictions were carried out and 81 % of them in fact proved to be possible in contrast to the prediction. This is called conservativeness: the threshold that distinguishes between a positive and a negative prediction more often than necessary warns the professional the chosen equipment is due to fail. In 81 % of these negative predictions though, practice shows this was not necessary. If the contractor would have chosen more powerful equipment, this probably would have meant higher costs for no reason.

Table 3: Imaginary completed contingency table to illustrate a 'conservative' sensitivity-pair

	Predictions		
	−	+	Total
Obs.			
−	TN = 30	FP = 4	O(−) = 34
+	FN = 131	TP = 87	O(+) = 218
Total	P(−) = 161	P(+) = 91	N = 252

What is the tolerance on the difference between the prediction and the observation to define a 'positive' observation? One sensitivity-pair shows little to zero tolerance on the difference between the prediction and the observation. In practice, the contractor will always incorporate a certain tolerance to count for all uncertainties in the building pit. This means that for a positive model prediction the contractor will always investigate the available equipment at that time and use this information in I the decision. Therefore a ROC-curve probably says more about a prediction model than a single sensitivity-pair. This paper however aims to introduce the concept of sensitivity-pairs and ROC-curves. The reader is challenged to elaborate on the matter described.

A contractor will not choose a solution that is likely to fail. The probability of failure in this type of work is large and usually it is hard to calculate the financial consequences. Instead the contractor will choose slightly over dimensioned equipment, rather than the 'quick and dirty' solution. The prize for over dimensioned equipment is expected to be much lower than the costs of delay due to malfunctioning equipment and an unsafe working environment.

RESULTS

All 252 projects in the GeoBrain experience database have been 'post'dicted, using the prediction tools described above. This resulted in six sensitivity-pairs, one for each tool with the standard threshold values, which have been plotted in the ROC-space below (Fig. 1). The diagonal straight line indicates the line of no discrimination. A design method with a marker below or at this line is practically worthless: one may as well 'throw a coin'. ["When throwing a coin several times, it can be expected to get half the positives and half the negatives correct; this yields the point (0.5, 0.5) in ROC space. If it guesses the positive class 90 % of the time, it can be expected to get 90 % of the positives correct, but its false positive rate will increase to 90 % as well, yielding (0.9,0.90) in ROC space", according to Fawcett (2006)].

Model Comparison in Current Situation

The ROC plot in Figs. 1 and 2 with the corresponding sensitivity pairs and metric distances in Table 4 show that the BBN (Δ) and Hypervib-I (+) score better than the EAU rule (*) and the other three design tools (o, x and $\Diamond$). This means adding professionals' experience to empirical rules improves those predictions. This is interesting, especially in those cases where there is no time for time-consuming numerical calculations. Remarkably, the EAU-rule is better than the CUR-rule and both Hypervib-I derivatives (van Baars' method and Azzouzis method, see also Sects. 3.2.3 and 3.2.4), although it does not contain any soil related parameters. Both Hypervib-I derivatives end up at the line of no discrimination. Perhaps the rules only are applicable to a specific subset of projects, because they were originally set up within a subset of the currently used variable space.

Table 4: Summary of the current and optimized threshold values for all models described here

Model	Current threshold	Current sensitivity pair	Current metric distance[a]	Optimized threshold	Optimized sensitivity pair	Optimized metric distance[a]
CUR	0.005 m	(0.74;0.49)	0.90[b]	0.0029 m	(0.91;0.97)	0.91[c]
AZZ	$F_{used}/F_c >1$	(0.65;0.65)	0.74	d	d	d
Baars	$F_{used}/F_c >1$	(0.74;0.73)	0.79	d	d	d
Hypervib-1	0.06 m/min	(0.65;0.74)	0.70	0.08 m/min	(0.44;0.62)	0.58
EAU	$F_{used}/F_c >1$	(0.71;0.74)	0.76	d	d	d
BBN	38 %	(0.59;0.74)	0.64	36 %	(0.56;0.72)	0.63

If as stated in Sect. 4.2 the TPR and FPR are not equally important, the metric distance should be determined by factoring the vertical and horizontal distance to the 'perfect model'. Since the costs of FP's are higher than for FN's, the horizontal distance to FPR = 0 is much more important than the vertical distance to the value of TPR = 1. In the ultimate case, the FPR is the dominating aspect and the first coordinate of the sensitivity pair determines the quality of the model, with low values for the best models. As shown in Table 4 this results in only minor changes in the ranking of the models.

[a]The metric distance to the 'perfect model' is calculated by $\sqrt{(FPR^2 + (1 - TPR)^2}$

[b]Below the line of no discrimination

[c]Above the line of no discrimination

[d]Not applied yet

Improving Threshold Values

Every design rule has its own threshold value to distinguish between positive and negative predictions. Making a ROC-curve for each rule provides the best threshold for each model. Figures 3, 4 and 5 show three ROC curves for the CUR, the BBN and the Hypervib-I model respectively.

CUR

The threshold value for the CUR prediction varied between 0 and 0.01 m, using steps of 1e−4 m. In the current situation the threshold is 0.005 m which leads to a sensitivity pair of (0.74,0.49). This pair ends up below the 'line of no discrimination', which indicates that currently throwing a coin might provide better results than using the CUR-rule for the prediction. One might suggest to use a 'reversed' version of the rule: choose your equipment such as to stay under the required 0.005 m of displacement amplitude. Physically however, this would not make sense. In the extreme 0 displacement amplitude could be obtained and then the whole idea of vibratory driving is gone.

The optimized sensitivity pair reads (0.91,0.97), for a threshold value of 0.0029 m. Now the pair end up at the better side of the line. In the physical context a threshold of 0.0029 m is less conservative than the current threshold. This corresponds to the expert opinion that 0.005 m is quite conservative.

BBN

The threshold value for the BBN prediction varied between 0 and 100 %, using steps of 1 %. In the current situation the threshold is 38 % which leads to a sensitivity pair of (0.59,0.74). The optimized sensitivity pair reads (0.56,0.72), for a threshold value of 36 %. This pair ends up marginally closer to the 'perfect situation' (0,1). Practically there is no reason to change the threshold value. The metric distances column in Table 4 shows this with distance 0.64 that changes into 0.63.

Hypervib-I

The threshold value for the Hypervib-I prediction varied between 0.01 m and 1.00 m per 60 s, using steps of 0.01 m per 60 s. In the current situation the threshold is 0.06 m/min which leads to a sensitivity pair of (0.65,0.74). The optimized sensitivity pair reads (0.44,0.62), for a threshold value of 0.08 m/min. This pair ends up closer to the 'perfect situation' as well. The metric distances column in Table 4 shows this with the distance 0.70 that changes into 0.58.

Van Baars, Azzouzi and EAU Model

The Van Baars-model, the Azzouzi-model and the EAU-model have not been optimized. Both the Van Baars-model and the Azzouzi-model are derivatives from the Hypervib I-model and therefore it seems more logical to extract new derivatives from the updated original.

Model Comparison in the Improved Situation

Table 4 provides an overview of the current and the optimized threshold values. In the current situation the Hypervib-I -model and the BBN-model perform the best, compared by the others. The prediction value of the Hypervib-I model improved, using a slightly different threshold. Using the improved threshold, the Hypervib-I model performs better than the BBN-model. The van Baars model, the Azzouzi model and the EAU model have not been optimized. Remarkable is the fact that the optimized CUR-rule, the Dutch rule of thumb provides a threshold value (0.0029 m) that is comparable with the German rule of thumb that each 30 kN vertical cyclic force is needed per 100 kg sheet pile. Since in the EAU model is directly related to the frequency, this might be a coincidence. Nevertheless it is recommended to include in further investigations, keeping in mind that both the EAU model and the CUR model might have a shared origin.

CONCLUSIONS

This paper introduces the method of 'Receiver Operating Characteristic' to determine the quality of a model and to be able to optimize parameters and variables in the model. 252 field observations were used to re-examine prediction methods for the minimum required vibration force, based on a selection of Dutch cases and to prove the ROC method works. The Operating Characteristic (ROC)-space is suitable for three purposes: 1) determining the quality of a model, 2) objectively comparing several models to each other, given certain assumptions and 3) for threshold optimization within a model.

A sensitivity-pair, created from a confusion matrix that was filled by 252 comparisons between observations and predictions provides a position in the ROC-space that indicates the quality of the model used to make the predictions. Figure 1 shows six sensitivity-pairs, providing a quality label for the six models described in this paper, relative to the 'perfect model'. The ROC-space from Fig. 1 also enables a comparison between the six models, using the metric distance to the perfect model as a ranking order. In this ranking the numerical Hypervib-I model performs the best, closely followed by the model with added expert knowledge. A positive side effect of the ROC method comprises the possibility of threshold and parameter optimization. Figures 3,4 and 5 show the performance of three models in a ROC-curve, providing sensitivity pairs that indicate the best threshold value for each model.

Conclusively, the Receiver Operating Characteristic (ROC)-space is suitable for the objective comparison of several design models. Using project information from the GeoBrain observations database it is possible to validate the codes and to attach a performance label to them, making it much easier for an engineer or designer to choose the right code. Depending on the position in the ROC-space a design code can be labeled 'conservative' or not. Furthermore, the ROC-curve technique enables engineers to optimize threshold values in their codes, that in turn leads to better predictions and thus safer and cheaper projects. It was to be expected that the more experience is added to the rules and models, the more reliable prediction models become. Figures 1 and 2 prove this is true. The model with added professionals' experience currently performs better than all other models and after improving the numerical model it performs nearly equally well.

Appendix 1

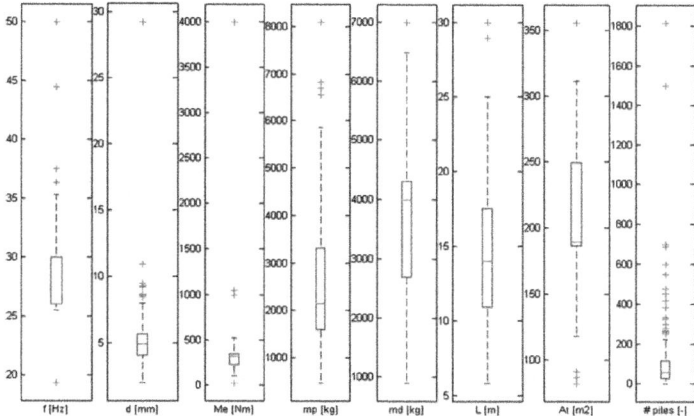

Figure 6: *Boxplot* of the main features from the Dutch projects used for the calculations.

REFERENCES

1. Azzouzi S (2003) Intrillen van stalen damwanden in niet-cohesieve gronden—welke predictie is (on)juist? (in Dutch). Masters thesis, Delft University of Technology

2. Barends FBJ (2005) Associating with advancing insight—Terzaghi Oration 2005. XVI international conference on soil mechanics and geotechnical engineering. Osaka, pp 217–248

3. Bles T, Al-Jibouri S, van den Adel J (2003) A risk model for pile foundations. ISARC 2003, 20th international symposium on automation and robotics in construction

4. CUR166 (2005) Damwandconstructies, 4e druk (in Dutch). (Civieltechnisch Centrum Uitvoering Research en Regelgeving)

5. Dierssens G (1994) Ein bodenmechanisches Modell zur Beschreibung des Vibratios-rammens in körnigen Böden. University of Karlsruhe

6. EAU (1990) Empfehlungen des Arbeitsausschusses 'Ufereinfassungen'— Häfen und Wasserstrassen. Ernst und Sohn

7. Fawcett T (2006) An introduction to ROC analysis. Pattern Recognit Lett 27(Special Issue on 'ROC' Analysis in Pattern Recognition):861–874

8. Gonin H, Holeyman A, Rocher-Lacoste F (eds) (2006) TRANSVIB 2006: vibratory pile driving and deep soil compaction. Laboratoire Central des Ponts et Chaussées, Paris. ISBN 2-7208-2466-6

9. Hemmen BR (2005) The synergy between theory and practice in geo-engineering. XVI international conference on soil mechanics and geotechnical engineering. Osaka, pp 2809–2811

10. Holeyman A, Legrand C (1994) Soil modeling for pile vibratory driving. U.S. FHWA "International Conference on Design and Construction of Deep Foundations", Orlando, Florida, December 1994, vol II, pp 1165–1178

11. Holeyman A, Legrand C, Van Rompaey D (1996) A method to predict the driveability of vibratory driven piles. 3rd international conference on the application of stress-wave theory to piles. Orlando, USA, pp 1101–1112

12. Holeyman A, Vanden Berghe J-F, De Cock S (1999) Model testing of vibratory driven piles, vol. 2. In: Proceedings of the XIth ECSMFE, Amsterdam, June 1999, pp 769–776

13. Holeyman A, Vanden Berghe J-F, Charue N (2002) Vibratory pile driving and deep soil compaction. Zwets & Zeitlinger, Lisse. ISBN 90 5809 521 5

14. Mens AMJ, van Tol AF (2010) Validating models against experience in foundation engineering, using the ROC curve. NUMGE 2010. Trondheim

15. Mens AMJ, van Tol AF, Koelewijn AR (2008) Optimizing foundation engineering, validating models against experience using artificial intelligence. In: Singh DN (ed) IACMAG. Mumbai, India, pp 3384–3391

16. Metz CE (1978) Basic principles of ROC analysis. Seminars in nuclear medicine, vol. VIII, No. 4 (October), pp 283–298

17. Mitchell TM (1997) Machine learning. McGraw-Hill, Singapore

18. Van Baars S (2004) Design of sheet pile installation by vibration. Geotech Geol Eng 22:391–400

19. Vanden Berghe J-F (2001) Sand strength degradation within the framework of vibratory pile driving. Faculty of Applied Science. Louvain: université catolique de Louvain

20. Viking (2002) Vibrodriveability—a field study of vibratory driven sheet piles in non-cohesive soils. Division of Soil and Rock Mechanics. Stockholm, Sweden: Royal Institute of Technology

CITATION

CHAPTER 1

Ma, J. (2015) Influence Analysis of a New Building to the Bridge Pile Foundation Construction. Open Journal of Civil Engineering, 5, 109-117. doi: 10.4236/ojce.2015.51011.

CHAPTER 2

Shinya Inazumi (2011). Hydraulic Conductivity of Steel Pipe Sheet Pile Cutoff Walls at Coastal Waste Landfill Sites, Integrated Waste Management - Volume I, Mr. Sunil Kumar (Ed.), ISBN: 978-953-307-469-6, InTech, DOI: 10.5772/17485.

CHAPTER 3

Ding, J. , Cao, Y. , Wang, W. , Zhao, T. and Feng, J. (2014) Experimental Study of Dynamic Characteristics on Composite Foundation with CFG Long Pile and Rammed Cement-Soil Short Pile. Open Journal of Civil Engineering, 4, 1-12. doi: 10.4236/ojce.2014.41001.

CHAPTER 4

W. Elsamee, "New Method for Prediction Pile Capacity Executed by Continuous Flight Auger (CFA)," Engineering, Vol. 5 No. 4, 2013, pp. 344-354. doi: 10.4236/eng.2013.54047.

CHAPTER 5

Qissab, M. (2015) Flexural Behavior of Laterally Loaded Tapered Piles in Cohesive Soils. Open Journal of Civil Engineering, 5, 29-38. doi: 10.4236/ojce.2015.51004.

CHAPTER 6

F. Hage Chehade, M. Sadek and D. Bachir, "Numerical Study of Piles Group under Seismic Loading in Frictnal Soil—Inclination Effect," Open Journal of Earthquake Research, Vol. 3 No. 1, 2014, pp. 15-21. doi: 10.4236/ojer.2014.31003.

CHAPTER 7

Yongjei Lee, Sungchil Lee, and Hun-Kyun Bae, "Design of Jetty Piles Using Artificial Neural Networks," The Scientific World Journal, vol. 2014, Article ID 405401, 12 pages, 2014. doi:10.1155/2014/405401

CHAPTER 8

B. R. Jayalekshmi, S. V. Jisha , R. Shivashankar, Wind load analysis of tall chimneys with piled raft foundation considering the flexibility of soil, DOI 10.1007/s40091-015-0085-6.

CHAPTER 9

C.G. Chinnaswamya & David N.G. Chew Chiata, Assessment of pile response due to deep excavation in close proximity—A case study based on DTL3 Tampines West Station, DOI:10.1080/23311916.2015.1014247.

CHAPTER 10

A. W. Stuedlein, S. C. Reddy and T. M. Evans, Interpretation of augered cast in place pile capacity using static loading tests, DOI:10.1179/1937525514Y.0000000003.

CHAPTER 11

Xiaolin Weng, Jianxun Chen, and Jun Wang, "Fiber Bragg Grating-Based Performance Monitoring of Piles Fiber in a Geotechnical Centrifugal Model Test," Advances in Materials Science and Engineering, vol. 2014, Article ID 659276, 8 pages, 2014. doi:10.1155/2014/659276.

CHAPTER 12

Jinxing Lai, Houquan Liu, Junling Qiu, and Jianxun Chen, "Settlement Analysis of Saturated Tailings Dam Treated by CFG Pile Composite Foundation," Advances in Materials Science and Engineering, vol. 2016, Article ID 7383762, 10 pages, 2016. doi:10.1155/2016/7383762.

CHAPTER 13

A. M. J. Mens, M. Korff, A. F. van Tol, Validating and Improving Models for Vibratory Installation of Steel Sheet Piles with Field Observations, DOI 10.1007/s10706-012-9506-5.

INDEX